PRINCIPLES OF PLANT AND ANIMAL PEST CONTROL

VOLUME 5

Vertebrate Pests: Problems and Control

SUBCOMMITTEE ON VERTEBRATE PESTS
COMMITTEE ON PLANT AND ANIMAL PESTS
AGRICULTURAL BOARD
NATIONAL RESEARCH COUNCIL

NATIONAL ACADEMY OF SCIENCES
WASHINGTON, D.C.   1970

This report is one of a series on principles of controlling pests and diseases of plants and animals. The following volumes are in the series:

Volume 1   Plant-Disease Development and Control
Volume 2   Weed Control
Volume 3   Insect-Pest Management and Control
Volume 4   Control of Plant-Parasitic Nematodes
Volume 5   Vertebrate Pests: Problems and Control
Volume 6   Effects of Pesticides on Fruit and Vegetable Physiology

The reports were prepared by six subcommittees working under the direction of the Committee on Plant and Animal Pests. The following organizations sponsored the work:

Agricultural Research Service, U.S. Department of Agriculture
Agency for International Development, U.S. Department of State
National Agricultural Chemicals Association
Rockefeller Foundation
Bureau of Sport Fisheries and Wildlife, U.S. Department of the Interior

ISBN 0-309-01697-5

Available from

Printing and Publishing Office
National Academy of Sciences
2101 Constitution Avenue
Washington, D.C. 20418

First printing, June 1970
Second printing, June 1971

Library of Congress Catalog Card Number 68-60085

Printed in the United States of America

Foreword

The objective of the project on Plant and Animal Pest Control was to outline, for each of the several classes of pests, the principles of control where these are established; to call attention to effective procedures where true principles are not yet established; and to indicate areas of research that appear to warrant early attention. The reports are not intended to be textbooks, in the usual sense, or encyclopedias, but they are intended to deal with basic problems, the principles involved in controlling pests, and the criteria that should be considered in conducting research and in evaluating published information. Specific instances of control practices are cited only to illustrate principles and procedures. It is hoped that these reports will be useful to researchers at all levels, to pest-control agencies, to administrators seeking guidance on priorities for application of resources, and to general field workers in the United States and elsewhere.

The National Academy of Sciences selected a committee of outstanding scientists to represent the diverse aspects of the problem and assigned to them responsibility for carrying out the project. To assist that committee, six subcommittees of specialists were appointed. Appropriate members of the parent committee were assigned as liaison members of the subcommittee, and in due time all reports were reviewed by the parent committee.

Some seventy scientists have collaborated over a four-year period to produce this series. Many others have contributed, to a lesser degree, in preparing statements and in reviewing and commenting on drafts of individual sections. Final responsibility for the content of these volumes rests with the parent committee. The Agricultural Board, under whose direction the Committee on Plant and Animal Pests operated, has reviewed and approved each manuscript.

Committee Members

CHARLES E. PALM, Cornell University, *Chairman*

WALTER W. DYKSTRA, Fish and Wildlife Service, U.S. Department of the Interior

GEORGE R. FERGUSON, Geigy Agricultural Chemicals

ROY HANSBERRY, Shell Development Company

WAYLAND J. HAYES, JR., Communicable Disease Center, U.S. Department of Health, Education, and Welfare

LLOYD W. HAZLETON, Hazleton Laboratories, Inc.

JAMES G. HORSFALL, Connecticut Agricultural Experiment Station

E. F. KNIPLING, Agricultural Research Service, U.S. Department of Agriculture

LYSLE D. LEACH, University of California, Davis

ROY L. LOVVORN, North Carolina State University

GUSTAV A. SWANSON, Colorado State University

Preface

From his very beginning, man has been in conflict with certain other vertebrates. He has tended to assume that the offending animals are wholly at fault and should be destroyed, failing to consider his responsibility for employing preventive or ecologically sound corrective measures. The Subcommittee on Vertebrate Pests was asked to examine, to assess, and to recommend on such man-vertebrate conflicts. Their deliberations are presented in this volume.

Pest control efforts are concerned primarily with limiting the entry and speeding the exodus of animals that are, or could be, inimical to the welfare of man. The techniques for accomplishing these ends differ among animal groups and among species within groups.

All programs for controlling the populations of animals, whether they limit birth rates or increase death rates, must strive to be specific for target species or individuals. They must be ecologically sound, socially acceptable, and harmless to the remaining biota, including man.

Many people respond emotionally to proposals to control animal populations, and groups seldom agree on what species should be controlled or on what control measures should be used. Any proposal for control may become controversial. We could not hope to reconcile all differences of opinion, but we trust that the material provided here will help to produce sound decisions.

The Introduction and the concluding chapter, Where We Stand, were prepared by Robert A. McCabe, Robert E. Lennon, and Edward L. Kozicky, who received suggestions and contributions from other subcommittee members. The remaining chapters and their respective authors are as follows:

Fishes in Pest Situations, Robert E. Lennon

Amphibians and Reptiles as Pests, John C. Neess

Birds in Pest Situations, Edward L. Kozicky and Robert A. McCabe
Small-Mammal Pest Situations, W. Robert Eadie and Nelson B. Kverno
Predatory Mammals That Become Pests, Robert A. McCabe and Milton
   Caroline
Pest Situations Involving Big Game, Richard D. Taber

SUBCOMMITTEE ON VERTEBRATE PESTS

ROBERT A. McCABE, *Chairman*, University of Wisconsin
MILTON CAROLINE, U.S. Department of the Interior
W. ROBERT EADIE, Cornell University
EDWARD L. KOZICKY, Olin Corp.
NELSON B. KVERNO, U.S. Department of the Interior
ROBERT E. LENNON, U.S. Department of the Interior
JOHN C. NEESS, University of Wisconsin
RICHARD D. TABER, Montana State University
ALEXANDER ZEISSIG, Cornell University
WALTER W. DYKSTRA, *ex officio*, Fish and Wildlife Service, U.S. Depart-
   ment of the Interior

Contents

*Vertebrate Pests:
Problems
and
Control*

CHAPTER 1

Introduction

Pest situations increasingly affect man's survival, and the trend will continue. Not only does population growth place man in more frequent contact with pest situations, but it has exacerbated man's struggle to feed himself. This report deals with vertebrate animals that confront man in pest situations.

We lack full information concerning the mechanical devices, repellents, and poisons with which we control vertebrate pests. Rapid advances in the chemical technology of poisons and repellents have outstripped our understanding of their side effects. *Silent Spring* (Carson, 1962) focused attention on these problems and precipitated a bitter debate that we shall not enter here. Rudd (1964) summarized the hazards of pesticide use, with supporting data when available. The scientific and lay communities are now fully alerted to the need for continuing scrutiny of our plant and animal control programs.

What is a pest situation? In brief, such a situation results when the activities of another vertebrate conflict with the interests or welfare of man: when sharks, singly or in packs, disrupt commercial fishing, destroy gear, or drive swimmers from water; when the alewife invades the upper Great Lakes, explodes to superabundance, clogs water intake lines, and, in annual die-offs, fouls shorelines and beaches; when the cane bullfrog (*Bufo marinus*), prolific, poisonous, and noisy, is introduced in Australia; when concentrations of birds are hazardous to aviation, destructive to grain crops, and serious nuisances in cities; and when large mammals threaten man's food, fiber, and recreational resources.

Pest situations range from those in which animals are annoying to man to those in which they prey on man. Midway in the range are situations in which they are in competition with man. The more technologically advanced societies tend to react quickly to pests, even when they are merely a nuisance. Primitive societies tend to react positively only when their very livelihood is threatened.

While pest situations may involve anywhere from a small to a large part of a given animal population, the resulting effects on man are not necessarily proportional. A small segment of a field mouse population or deer population, for example, can cause extensive damage to commercial orchards, but large numbers of house sparrows or pet dogs cause relatively little discomfort or economic loss in cities. A serious pest situation in one set of circumstances may well be tolerable in another.

Pest situations become complicated when they bring groups of people into conflict—when differences of opinion arise about the seriousness of a pest's activities or about the appropriateness of proposed control measures. Here economic values confront aesthetic considerations and value judgments must be reached. The makers, sellers, and users of control materials plead for maximum control to increase crop yields and raise the standard of living; the conservationist pleads for minimum or no control; the ecologist pleads for the resource and its environment as basic to both economy and preservation.

No organisms encountered as pests stir as much emotion as do the vertebrates, particularly birds and big game. The preservationist attitude is compounded where aesthetically desirable fishes, birds, and mammals are concerned.

Conflicts that involve the public are usually resolved by legislative action. Lawmakers often lack professional training or background in these areas but must nevertheless render important decisions that have biological implications. And because of the wide diversity of pest situations, legal responsibilities are often poorly defined.

Concise legal statements are needed to define, clarify, and promote the proper course of action when an otherwise beneficial vertebrate animal becomes a pest. Although laws pertaining to animal control should be flexible enough to cover the diverse nature of pest problems, they should not allow emotional public reaction to vitiate considered solutions of those problems. The development of a model law would help bring about public acceptance of considered solutions.

The control of vertebrate pests is of interest to many groups, such as farmers, merchants, foresters, sportsmen, industrialists, and ecologists. Each views the problem from its own vantage point, and they have often worked against each other. For many years, the strongest cohesive force was exerted by purveyors of control devices, chemicals, and services. As pest problems spread, forums were organized to facilitate exchange of ideas and coordination of control efforts:

• Meetings were sponsored by the National Pest Control Association (NPCA), whose members are professional pest control operators. These meetings were devoted largely to control procedures, the economics of control, and legislative proposals for limiting the use of pesticides.

● In February 1962, the California Pest Control Technical Committee of the NPCA called a conference of agencies and individuals interested in ". . . controlling troublesome and pestiferous birds, mammals and snakes, and to discuss related problems concerning diseases and pesticides" (Howard, 1962). More than 300 persons attended. Similar conferences were held in 1964 and 1967. Proceedings of the three conferences were published by NPCA.

● In 1962, 1964, 1966, and 1968, seminars on bird control were held at Bowling Green, Ohio, under the sponsorship of the Bowling Green State University, the U.S. Fish and Wildlife Service, and the NPCA. Proceedings of two of the seminars have been published (Beck and Jackson, 1964; Schneider and Jackson, 1966).

● In 1959, The Wildlife Society established a Committee on Economic Losses Caused by Vertebrates. The committee had these objectives: (1) to "encourage studies within the profession to measure objectively economic losses caused by vertebrates"; (2) to "encourage exchange of information between the many groups interested in vertebrates and members of The Wildlife Society which will lead to an understanding at a high level of the problems involved in coping with economic losses by vertebrates"; and (3) to "devise ways of keeping the wildlife profession informed on the subject so they will exercise good judgment in their own field of research and management work and especially in dealing with other interested groups."

● The Committee on Bird Protection of the American Ornithologists' Union has indicated its awareness of the need for bird control. The 1957 committee reported on collisions of albatrosses with aircraft (Kalmbach *et al.*, 1958). The 1964 report considered collisions between birds and aircraft, grain depredations, and methods of control. The 1965 report included an article on "Too Many Birds."

● The Conservation Committee of the Wilson Ornithological Society in recent years has devoted a section of its reports to pest problems under the caption *Control of Bird Populations.*

At its annual meeting in 1956, the American Fisheries Society passed a resolution requesting the U.S. Fish and Wildlife Service to expand its basic research into manipulation of fish populations and improvement of game-fish brood stocks through chemical treatment of waters. In 1958 the Society commended the Fish and Wildlife Service for expanding the program of testing selective toxicants for various species of undesirable fishes, and urged that high priority and adequate funds be given to the research. Shortly thereafter, the Service established the Fish Control Laboratories at La Crosse, Wisconsin, and Warm Springs, Georgia.

The American Society of Ichthyologists and Herpetologists has reacted strongly to the increasing use of general toxicants for mass eradication of fish and other organisms from large bodies of water. Its Committee on Fish Conservation has endeavored to convince federal and state agencies that the anticipated benefits of eradication programs—such as temporary improvements in sport fisheries—must be weighed against the possible deleterious effects on unique species, and it criticized the poisoning of fish throughout more than 500 miles of the Green River system in Wyoming, Utah, and Colorado in 1962 (Hubbs, 1963). As a result, the Secretary of the Interior issued directives regarding fishery management projects that will "... assure that there is no damage to the environment beyond the intended scope of the operation." Thus, the Society has been instrumental in protecting nontarget and unique organisms.

The Smithsonian Institution, in cooperation with the Office of Naval Research, has maintained a shark attack file since 1959. Worldwide data on attacks on bathers and on victims of sea and air disasters are gathered and analyzed (Schultz, 1967).

On the whole, however, response by scientific societies to the need for rational inquiry into pest situations and their alleviation has tended to be lacking, intermittent, or restricted in scope.

Citizens' organizations interested in the welfare and ecological behavior of vertebrates are becoming increasingly aware of the need for vertebrate pest control.

On May 16, 1967, the Secretary of the Interior announced a policy on wildlife control that had been reviewed by 30 conservation organizations and agencies, including the Secretary's Advisory Board on Wildlife Management. The objectives were to protect human health by suppressing wild animals that are vectors of disease; to control birds that may cause damage to airplanes in the vicinity of airports; to protect urban areas from damage by rats and mice; to protect forests and rangeland from animals that hinder restoration; to protect growing crops and stored products from damage by birds and rodents; to protect livestock from depredation; and to suppress wild animals that are vectors of livestock diseases. The following were among the statements of policy:

- Whether wild animals are beneficial or injurious depends on time and place.
- The control of wildlife populations as a resource is necessary along with land acquisition, habitat manipulation, and harvest regulation.
- The bounty system will not be used.

Another response to the need to control wildlife is the appointment of an impressive task force by the Secretary of the Interior to study the alewife, a recent invader of the upper Great Lakes, where it has become extremely abundant and has clogged water intake systems and littered beaches in huge annual die-offs. The task force will evaluate ways to alleviate the situation, such as harvesting alewives for fish meal, stocking the lakes with alewife predators (e.g., lake trout and coho salmon), improving methods for collecting dead alewives before they reach the shores, and cooperative cleanup campaigns of alewife-polluted areas.

REFERENCES

Beck, J. R., and W. B. Jackson (eds.). 1964. Proc. 2nd Bird Control Seminar. Bowling Green State Univ. 140 pp.

Carson, R. 1962. Silent Spring. Houghton Mifflin Co., Boston. 368 pp.

Howard, W. E. 1962. Vertebrate pest control, pp. 1–7. *In* Proc. Vert. Pest Control Conf., Nat. Pest Control Ass., Elizabeth, N.J. 391 pp.

Hubbs, C. L. 1963. Secretary Udall reviews the Green River fish eradication program. Copeia 2:465–466.

Kalmbach, E. R. (Chairman), Jean Delacour, Ira N. Gabrielson, Robert A. McCabe, and David A. Munro. 1958. Report to the American Ornithologists' Union of the Committee on Bird Protection, 1957. The Auk 75(1):81–85.

Rudd, R. L. 1964. Pesticides and the living landscape. Univ. of Wisconsin Press, Madison. 320 pp.

Schneider, D. E., and W. B. Jackson (eds.). 1966. Proc. 3rd Bird Control Seminar. Bowling Green State Univ. 172 pp.

Schultz, L. P. 1967. Predation of sharks on man. Chesapeake Sci. 8(1):52–56.

CHAPTER 2

Fishes in Pest Situations

Man's reliance on fish as food is indicated in the fact that in 1965 the world catch amounted to 111 billion pounds. Even so, the oceans, which cover 71 percent of the earth's surface, comprise the remaining largely undeveloped food source. The expected human population explosion will no doubt greatly increase the worldwide demand for fish to alleviate food shortages.

Man's long-standing association with freshwater and marine fishes has not been without serious problems. Pest situations in fact and fiction abound in the literature. Since some fish grow to great size, they have given rise to myths and legends of monsters. For example, the oarfish, a marine species that may grow to 50 feet in length, has bright scarlet fins and an elongated dorsal fin, and a head somewhat resembling that of a horse. When viewed in the water under certain conditions, it could easily be considered a monster or sea serpent.

Though the legends may be incredible, there are truly some bizarre problems with fish today that affect man's safety and welfare. The following accounts merely touch on some of the more important pest situations involving freshwater and marine fishes.

FISHES THAT AFFECT MAN'S SAFETY

Fishes that affect man's safety include the biters, electric fish, venomous fish, poisonous fish, and fishes that serve as vectors for helminthoses. They cause extensive trouble in fresh waters and oceans around the world. Some difficulties, such as with poisonous fishes, are of considerable importance from a military standpoint and have warranted intensive investigations by the United States, Japan, the Soviet Union, and other countries (Halstead, 1965).

BITING FISHES (TABLE 1)

Sharks

Man has long feared sharks. Their ability to maim and kill is undisputed. Among them, the black-tipped shark, hammerhead, Lake Nicaragua, sand tiger, tiger, and white shark are most dangerous, but 24 other kinds are potentially dangerous. About 75 attacks are recorded each year, but in remote regions many attacks may not be recorded (Schultz, 1967). Although the number of fatalities and severe injuries is not great, the danger of an attack is widespread in open oceans, off beaches, and in river estuaries in tropical and temperate regions. An analysis of 1,406 attacks on humans disclosed that most took place where there were concentrations of bathers, usually within a thousand feet of the shore (Schultz, 1967). Typically these incidents involve single sharks, but schools of predaceous sharks may attend sea and air disasters and may cause many deaths. The U.S. Office of Naval Research maintains a shark attack file at the Smithsonian Institution, Washington, D.C.

The gruesome nature rather than the incidence of shark attacks stimulates interest in control measures. However, relatively little control in reducing attacks is possible today (see Table 8). Shark fences and shark patrols are used at some Australian beaches to prevent attacks on bathers (Ommanney, 1963); shark repellents have been failures to date. Fighting back may deter an attacking shark: one investigator found that 61 percent of attacking sharks were repelled by hitting or stabbing them, but 39 percent continued their attacks (Schultz, 1967). Cessation of the practice of dumping garbage, offal, and other refuse at sea near beaches may reduce concentrations of sharks potentially dangerous to bathers.

Piranhas

The vicious piranhas (caribes) are the most feared fish in the Orinoco and Amazon basins in South America. These fish are seldom more than 18 in. long, but they possess extraordinarily powerful and sharp dentition. They travel in large schools and are able to overwhelm and quickly skeletonize large prey, including man and domestic animals. They are especially dangerous away from the main currents of rivers, such as in calm, shallow waters or estuaries. Fontenele (1963) noted that they were becoming abundant and troublesome in new hydroelectric impoundments until rotenone was employed as a toxicant to bring them under control.

Fear of injury or death keeps most people out of piranha-infested waters. Stations are set up in particularly dangerous districts to warn of the arrival of piranha schools. More positive controls may have to be exerted as civilization moves into infested districts.

TABLE 1 Pest Situations Involving Biting Fishes and Human Safety

Species	Problem		Location	Season	Reference
	Nature	Loss			
SHARKS					
Blacktip	Man-eater	Injury, death	Cuba, Florida	All year	Bigelow and Schroeder, 1948
Hammerhead	Man-eater	Injury, death	Tropical, temperate seas	All year	Bigelow and Schroeder, 1953
Lake Nicaragua	Man-eater	Injury, death	Nicaragua	All year	Bigelow and Schroeder, 1953
Sand tiger	Man-eater	Injury, death	Tropical, temperate seas	All year	Hildebrand and Schroeder, 1928
Tiger	Man-eater	Injury, death	Tropical, temperate seas	All year	La Gorce, 1952
White	Man-eater	Injury, death	Tropical, temperate seas	All year	Ommanney, 1963
CHARACINS					
Piranhas	Man-eater	Injury, death	South American rivers	All year	Caras, 1964
MURAENIDS					
Moray eel	Biter	Injury	Tropical seas	All year	Grosvenor, 1965
Serpent eel	Biter	Injury	Adriatic Sea	All year	Soljan, 1963
PYGIDIIDS					
Candiru	Invades genitalia	Injury, infection	Amazon River system	All year	Lagler, Bardach, and Miller, 1962
SPHYRAENIDS					
Great barracuda	Biter	Injury	Tropical, subtropical seas, West Indies	All year	Norman, 1963

Eels

Moray eels inhabit temperate and tropical coastlines and are vicious biters. Most are 4–5 ft long, but there are records of giants up to 10 ft long (Herald, 1961). Although they are a widespread and constant hazard to fishermen and swimmers, they tend to keep out of sight, and few are seen even where their population is large. No controls other than fishing and spearing are practiced.

Candiru

Another freshwater fish greatly feared in South America is the candiru (carnero), a diminutive catfish of the genus *Vandellia*, which penetrates the urogenital opening of swimming mammals and man (Lagler, Bardach, and Miller, 1962). About one inch long, the candiru has an elongated, compressed, and translucent body. It is reported to be attracted by a discharge of urine in the water, and once it has penetrated the vagina or penis, it becomes wedged by erectile spines on its gill covers, causing severe inflammation and hemmorhage. Various shields made of coconut shells or palm fibers are employed by native bathers to protect genitalia from invasion by the fish (Gudger, 1930). No positive means for controlling candiru in remote, tropical waters has been reported.

Great Barracuda

Circumtropical in distribution, the great barracuda is sometimes feared more than sharks and is most dangerous to man in the vicinity of the West Indies. There are records of many attacks on swimmers by great barracuda 6 ft or more in length. An attack usually consists of a single strike, which leaves a large, clean wound. Large barracuda tend to remain alone and to range rather widely; thus, positive controls are impractical. Care in recognizing the fish and avoiding attack is advisable in barracuda-infested waters.

ELECTRIC FISHES (TABLE 2)

Electric Rays

Electric shocks of up to 200 volts have been measured in rays, but the average is 70 to 80 volts. A fisherman or bather stepping on an electric ray may be merely tickled or may be knocked down by the shock, depending on the strength of the discharge. These fish have been known since antiquity, and some were credited with therapeutic powers in Greek and Roman literature (Herald, 1961; Norman, 1963). No positive controls have been reported.

TABLE 2 Pest Situations Involving Electric Fishes and Human Safety

Species	Problem		Location	Season	Reference
	Nature	Loss			
TORPEDINIDS					
Common torpedo	Electric shock	Injury	Tropical, temperate seas	All year	Bigelow and Schroeder, 1953
Pacific torpedo	Electric shock	Injury	Tropical, temperate seas	All year	La Gorce, 1952
GYMNOTID					
Electric eel	Electric shock	Injury	Amazon and other rivers, northeast South America	All year	Norman, 1963

Electric Eel

The electric eel, which is not a true eel but a gymnotid, is found in South American rivers. Ranging up to 6 ft long, it can produce as much as 600 volts, but the average maximum is about 350 volts. Many stories are told of men and pack-beasts being knocked down by shocks from the fish while fording rivers.

One wonders what the response of the electric eel might be to control toxicants that are taken in across the gills. The eel is largely an air breather and the gills are not used much in respiration. If control becomes desirable, some extraordinary method may have to be found for this unusual fish.

VENOMOUS FISHES (TABLE 3)

There is a wide variety of freshwater and marine fishes that have venomous spines; their hazards to man are widespread and serious.

The greater weever (derived from an old Anglo-Saxon word meaning viper) of Europe, extending from the Black Sea to the Baltic, is greatly feared as the most poisonous of European fish (Nikol'skii, 1961). Although no more than 17 in. long, the weever has a poison spine in each gill cover and five or six spines in the first dorsal fin. The venom is both a blood and nerve poison, and severe injuries and death have been attributed to it (Halstead, 1965).

The venoms of stingrays are powerful and dangerous to waders and fishermen, and death may result from contact. Natives of the Indo-Pacific region have long used the spines of stingrays as weapons with great effect (Herald, 1961). Also dangerous and sometimes deadly are the scorpionfishes of subarctic, temperate, and tropical seas. Waders, divers, or fishermen run great risk if they step on or grasp such a fish, because each dorsal fin spine is grooved to carry venom from the poison glands at the base of the fin. The stonefish, an ugly species that resembles a small rock when lying partly buried on the ocean floor, possesses the most deadly of all fish venoms (Herald, 1961).

No practical controls are known for venomous fishes. Fishermen, recreationists, and investigators are advised to recognize the fishes and their habitats and to exercise every precaution to avoid contacts.

POISONOUS FISHES (TABLE 4)

Many species of fish are poisonous to eat. Some are always poisonous and others only seasonally so. Also, a species may be consistently toxic in one area but safe and valued as a food fish in a nearby area. Each year fish poisoning affects from a few hundred to perhaps thousands of people, and it kills a

TABLE 3 Pest Situations Involving Venomous Fishes and Human Safety

Species	Problem		Location	Season	Reference
	Nature	Loss			
RAJIDS					
Stingray	Poison spines	Injury, death	Tropical, subtropical seas	All year	Halstead and Mitchell, 1963
TRACHINIDS					
Weeverfishes	Poison spines	Injury, death	Northeast, southeast Atlantic; southeast Pacific oceans	All year	Norman, 1963
CHIMAEROIDS					
Rabbitfishes	Poison spines	Injury	Tropical Indo-Pacific	All year	Herald, 1961
SILURIDS					
Catfishes	Poison spines	Injury	Worldwide; freshwater and marine	All year	Halstead and Mitchell, 1963
SCORPAENIDS					
Zebra fishes	Poison spines	Injury, death	Indo-Pacific, Red Sea	All year	Halstead and Mitchell, 1963
Scorpionfish	Poison spines	Injury, death	California coast	All year	La Gorce, 1952
Stonefish	Poison spines	Injury, death	Indo-Pacific, Red Sea	All year	Herald, 1961
BATRACHOIDIDS					
Toadfishes	Poison spines	Injury	Tropical, subtropical seas	All year	Halstead and Mitchell, 1963
ACANTHURIDS					
Surgeonfishes	Poison spines	Injury	Tropical, temperate seas	All year	Halstead and Mitchell, 1963

small percentage of them (Walford, 1958). Recent biotoxicological surveys indicate that 29 to 75 percent of the shore fishes of some tropical Pacific islands are toxic. This situation has provided grounds for long-standing food taboos and has been a great deterrent to the development of commercial fisheries (Halstead, 1962). It is a problem that must be solved before the abundant edible shore and reef fishes in tropical seas can be exploited to feed hungry peoples.

There are four broad classifications of fish poisons: Ichthyosarcotoxism is caused by eating the flesh of poison fish, ichthyoötoxism by eating poisonous roe, ichthyohemotoxism by consuming poisonous fish blood, and scombroid dermatitis by handling certain fish (see Table 4).

Certainly the outstanding poisonous fish are the puffers because they are widespread and are always dangerous; they are also considered to be a delicacy. There are about ninety species in warm and temperate seas, with maximum size about 3 ft. The organs and sometimes the flesh contain tetrodotoxin, a poison that kills about 60 percent of the persons affected. One species, the Japanese puffer (*fugu*), has often been used in suicides. Some nations inspect and license commercial fisheries and eating places that deal with puffers in order to reduce the risk.

No positive controls are known for poisonous fishes, but they may be demanded soon by increasing human populations. Taboos undoubtedly keep many people from being poisoned, and some taboos would be extremely difficult to eliminate even if positive controls made local fish safe to eat. In the literature there is frequent mention of suspect fish being first fed to cats or to swine to determine its safety. The Japanese military authorities published and distributed an excellent handbook on poisonous fishes to troops in the South Pacific during World War II. Other nations engaged in that area without such information soon recognized the need for it (Halstead, 1965).

FISHES AS HOSTS FOR HUMAN PARASITES (TABLE 5)

A large number of freshwater fishes, including some of the finest food and game fishes, may serve as intermediate hosts for dangerous parasites of man. These parasites include the fish tapeworm, intestinal flukes, and liver flukes. Where the incidence of infestation in fish is high, the threat to the health of humans constitutes a pest situation. Although certain public health measures, specified later, can help to alleviate the problem, the control of infected fish populations may be the most direct and effective solution in some circumstances. Detailed studies on fish as vectors for human parasites have been published by Bauer (1961) and Van Duijn (1962).

TABLE 4 Pest Situations Involving Poisonous Fishes and Human Safety

Species	Problem		Location	Season	Reference
	Nature	Loss			
Lampreys, hagfish	Ichthyosarcotoxism	Sickness, death	Tropical, temperate seas	All year	Ghiretti and Rocca, 1963
Sharks	Ichthyosarcotoxism	Sickness, death	Tropical seas	All year	Marshall, 1966
Rays	Ichthyosarcotoxism	Sickness, death	Tropical seas	All year	Walford, 1958
Chimaeras	Ichthyosarcotoxism	Sickness, death	Tropical seas	All year	Walford, 1958
Moray eels	Ichthyosarcotoxism	Sickness, death	Tropical seas	All year	Walford, 1958
Sea bass	Ichthyosarcotoxism	Sickness, death	Tropical seas	All year	Walford, 1958
Barracuda	Ichthyosarcotoxism	Sickness, death	Tropical seas	All year	Walford, 1958
Wrasses	Ichthyosarcotoxism	Sickness, death	Tropical seas	All year	Walford, 1958
Snappers	Ichthyosarcotoxism	Sickness, death	Tropical seas	All year	Walford, 1958
Parrotfish	Ichthyosarcotoxism	Sickness, death	Tropical seas	All year	Walford, 1958
Puffers	Ichthyosarcotoxism	Death, 100 per year	Tropical seas	All year	Halstead, 1965
Porcupinefish	Ichthyosarcotoxism	Sickness, death	Tropical seas	All year	Halstead, 1965
Tuna, mackerel	Ichthyosarcotoxism	Sickness, death	Tropical seas	All year	Halstead, 1965
Freshwater bream	Ichthyoötoxism	Sickness, death	Asia, Europe	Near spawning time	Halstead, 1965
Barbels	Ichthyoötoxism	Sickness, death	Asia, Europe	Near spawning time	Halstead, 1965
Fine-scale carp	Ichthyoötoxism	Sickness, death	Asia, Europe	Near spawning time	Halstead, 1965
Moray eels	Ichthyohemotoxism	Sickness	Tropical seas	All year	Halstead, 1965
Eels	Ichthyohemotoxism	Sickness	Subtropical, temperate seas	All year	Halstead, 1965
Tuna	Scombroid dermatitis	Dermatitis	Tropical, subtropical seas	All year	Halstead, 1962
Skipjack	Scombroid dermatitis	Dermatitis	Tropical, subtropical seas	All year	Halstead, 1962
Bonitos	Scombroid dermatitis	Dermatitis	Tropical, subtropical seas	All year	Halstead, 1962

TABLE 5 Pest Situations Involving Fishes as Intermediate Hosts for Human Parasites

Species	Disease	Location	Season	Reference
Salmons (six species)	Diphyllobothriasis	Japan	All year	Belding, 1942
Lake trout	Diphyllobothriasis	Europe, Asia, North America	All year	Bauer, 1961
Northern pike	Diphyllobothriasis	Europe, Asia, North America	All year	Bauer, 1961
Ruff	Diphyllobothriasis	Europe, Asia, Africa	All year	Oppenheimer, 1962
Barbel	Diphyllobothriasis	Europe, Asia, Africa	All year	Van Duijn, 1962
European eel	Diphyllobothriasis	Europe, Asia	All year	Bauer, 1961
Common perch	Diphyllobothriasis	Europe, Asia	All year	Bauer, 1961
Walleye	Diphyllobothriasis	Europe, Asia	All year	Bauer, 1961
Burbot	Diphyllobothriasis	Europe, Asia	All year	Bauer, 1961
White amur	Clonorchiosis	Russia, southeast Asia	All year	Bauer, 1961
Striped mullet	Clonorchiosis	Japan, China	All year	Belding, 1942
Carp	Opisthorchiasis	Eastern Europe, Asia	All year	Bauer, 1961
Roach	Opisthorchiasis	Eastern Europe, Asia	All year	Bauer, 1961
Tench	Opisthorchiasis	Eastern Europe, Asia	All year	Bauer, 1961
Bream	Opisthorchiasis	Eastern Europe, Asia	All year	Bauer, 1961
Rudd	Opisthorchiasis	Eastern Europe, Asia	All year	Bauer, 1961
Ide	Opisthorchiasis	Eastern Europe, Asia	All year	Bauer, 1961
Silver bream	Opisthorchiasis	Eastern Europe, Asia	All year	Bauer, 1961
European dace	Opisthorchiasis	Eastern Europe, Asia	All year	Bauer, 1961
Rudd	Metagonimiasis	Europe, Asia, Malaysia	All year	Bauer, 1961
Striped mullet	Heterophyiasis	Africa, Asia, Malaysia	All year	Belding, 1942
Freshwater mullet	Heterophyiasis	Hawaii	All year	Van Duijn, 1962

Diphyllobothriasis

This disease is caused by the broad tapeworm, *Diphyllobothrium latum*, and is contracted by eating raw fish or fish that has been inadequately cooked, salted, or smoked. The geographical center for infection in humans is around the Gulf of Finland where the rate of incidence ranges as high as 20 percent of the population (Van Duijn, 1962). It also occurs in the vicinity of Lake Superior and in Alaska. In one region in southwest Alaska, an incidence of 30 percent was recorded among Eskimos.

In a survey of numerous fish from the Gulf of Finland, the plerocercoids of *D. latum* were found in 88 percent of the northern pike, 91.6 percent of the burbot, 53 percent of the common perch, and 98 percent of the ruff (Bauer, 1961). In other waters of northwest USSR, plerocercoids were found in all Atlantic salmon and northern pike. Thus, the potential for human infection is very great.

Several public health measures have been proposed to help solve the tapeworm problem (Bauer, 1961). They are as follows: freshly salted fish should not be eaten, because the plerocercoids are not immediately killed by salt; the sale of uncooked roe of northern pike should be prevented; steps should be taken to eliminate the adult parasites in man; the discharge of untreated feces of man and domestic carnivores into waters should be prohibited, especially from ships and barges; and the long-standing custom of eating raw or partially cooked fish should be discouraged. The latter is a difficult task because raw fish is a favorite food in the northern USSR.

Clonorchiosis

The fluke *Clonorchis sinensis* parasitizes fish and the bile duct of man and carnivores. It is widespread in southeast Asia and only recently was discovered in the USSR (Bauer, 1961). The disease is serious, and its incidence can be curtailed only by complete processing of fish (salting, smoking, and dry curing) in a commercial plant or by storing at -10°C for at least three weeks. Strict sanitation must be observed also at animal and fur farms where raw fish is fed.

Other Flukes

Opisthorchiasis is caused by the fluke *Opisthorchis felineus*, commonly called the cat liver fluke. The disease is common in eastern Europe and Asia in the cat and man, and mass infestations can be fatal. The metacercariae, for example, are very common in the five major food fishes of the Dnieper River (Van Duijn, 1962). No effective treatment is known for the disease, and abstention from eating any but well-cooked fish is the best preventive measure.

Metagonimiasis is an intestinal disease caused by the fluke *Metagonimus yokogawai*, and it occurs in man as a result of eating insufficiently cooked or salted fish. Heterophyiasis is also an intestinal, parasitic disease of man and is caused by *Heterophyes heterophyes*, a fluke. The natural hosts are dogs, cats, and fur-bearing carnivores, to which it is transmitted through uncooked or insufficiently cooked fish.

The possibility of controlling helminthoses in fish cultural ponds and other small bodies of water by applying poisons to kill snails and fish should be explored. Improving sanitary practices to prevent raw fecal contamination and then restocking the water with parasite-free fish may be effective in reducing the tapeworm and fluke diseases in some places. Without these precautions, the increasing attempts to promote greater production and utilization of fish to meet food shortage could result in serious spread of the parasitic diseases.

FISHES THAT AFFECT MAN'S WELFARE

There are many species of freshwater and saltwater fishes that occasionally come into conflict with man's welfare. These conflicts occur in both standing and flowing waters, in natural and impounded waters, and in cultivated waters, and they are worldwide in scope. The pest situations—often resulting in great but unspecified economic or aesthetic losses—include destruction of commercial fishing gear; interruption of commercial or sport fishing; fouling of shorelines with dead fish; damage to fishery or waterfowl habitat; predation on shrimp or waterfowl; clogging of municipal water-intake lines; excessive competition with or predation on more valued food or game fish; and transmission of fish diseases.

MARINE FISHES (TABLE 6)

Spiny Dogfish

This small shark, 2 to 3 ft long, is widespread and extremely abundant in the North Atlantic and North Pacific Oceans. It travels in huge packs and is voracious almost beyond belief, probably causing more destruction to fishing gear and fishing operations than any other species of fish. When locally abundant in summer, the dogfish may make commercial and sport fishing completely futile. Nets and lines are badly damaged, catches in the nets or on lines are eaten, and shoaling food and game fishes are harried and driven from fishing grounds.

TABLE 6 Pest Situations Involving Marine Fishes and Human Welfare

Species	Problem		Loss	Location	Season	Reference
	Nature					
Hagfish	Destroys catch		Economic	Temperate seas	All year	Bigelow and Schroeder, 1953
Spiny dogfish	Destroys catch, damages gear		Economic	Temperate seas	Summer	Jensen, 1966
Soupfin, silky, and white-tipped sharks	Devour catch, damage gear, harry fish		Economic	Tropical and temperate seas	All year	Thompson and Springer, 1961
Basking shark	Damages drift nets		Economic	North Pacific, British Isles	Summer	U.S. Bureau of Commercial Fisheries, 1955
Indian tarpon	Predation in shrimp ponds		Economic	Taiwan	All year	Tang, 1961
Striped mullet	Predation in shrimp ponds		Economic	Taiwan	All year	Tang, 1961

18

The spiny dogfish is quite palatable, but repeated efforts to promote it as a food fish have met with little success. About two million pounds are landed annually in the United States, mostly for reduction into meal and oil. Positive management of the species is needed, and the discovery of a worthwhile and economically attractive use would be desirable. Otherwise, some form of biological control may be needed to reduce the damage caused to more valuable fish (Jensen, 1966).

Other Sharks

Many of the 300 species of sharks are at times involved in conflicts with man, and the number and severity of conflicts may be expected to increase as man makes more use of the seas for food and recreation. Sharks in the Gulf of Mexico often cause severe economic losses to commercial fisheries by biting and tearing trawl nets to get at the catch (Thompson and Springer, 1961). On a worldwide basis, sharks reportedly destroy about 20 percent of the saleable tuna taken in longline fisheries.

Commercial exploitation of sharks is small and apparently insufficient to reduce numbers. Some shark meat is used by humans, and some is reduced to meal. Hides are used for quality leather products. The principal exploitation was for shark-liver oil, which is rich in vitamin A, but with the advent of synthesized vitamin A this has declined sharply.

FRESHWATER FISHES (TABLE 7)

The story of the sea lamprey—its invasion of the Great Lakes in North America, its destruction of highly valued food and game fish, and its eventual control by chemical and electrical means—has been well documented and has received wide publicity in the past two decades. It serves as an excellent example of the definition of a pest situation and the ensuing development of selective controls through intensive research. Details are presented in another section of this report (p. 30).

Alewives

Normally an anadromous fish common in the North Atlantic, the alewife, like the sea lamprey, gained access to Lakes Erie, Huron, and Michigan through the Welland Ship Canal. After the lamprey decimated the predatory fish, the alewife exploded to superabundance. In 1965 it was estimated to constitute 90 percent of the weight of all fish in Lakes Huron and Michigan. Spawning groups moving inshore were of such size and density that they displaced even the tenacious yellow perch from their grounds (Moffett, 1966). In the past several

TABLE 7 Pest Situations Involving Freshwater Fishes and Human Welfare

Species	Problem		Location	Season	Reference
	Nature	Loss			
Sea lamprey	Predation on food fish	$8 million per year	Upper Great Lakes	All year	Baldwin, 1968
	Predation on food fish	Economic	Chesapeake Bay	Winter	Mansueti, 1962
	Predation on food fish	Salmon	Maine	Spring	Davis, 1967
Longnose gar	Predation	Game fish	Eastern, southwestern U.S.	All year	Burr, 1931
Shortnose gar	Predation	Game fish	Eastern, southwestern U.S.	All year	Randolph, 1965
Alligator gar	Predation	Game fish	Southern U.S.	All year	Caldwell, 1912
Bowfin	Predation	Game fish	Northeastern, southeastern U.S.	All year	Randolph, 1965
Alewife	Abundance, die-offs	Economic, nuisance	Great Lakes	Spring	Moffett, 1966
	Clogs intake lines	Nuisance	Chicago, New York	Spring	Gudger, 1950
Gizzard shad	Abundance, competition	Game fish	Oklahoma	All year	Jenkins, 1957
Brown trout	Competition	Game fish	California	All year	Nielson, 1953
Dolly Varden trout	Predation	Salmon	Western Canada, Alaska	Summer	Lagler and Wright, 1962
Brook trout	Disease carrier	Hatchery fish	Canada	All year	Armstrong, 1949
Chain pickerel	Predation	Trout	Maine	All year	Foye, 1964
Northern pike	Predation	Trout	Canada	All year	Smith, 1950
	Predation	Ducklings	Canada	Summer	Solman, 1945
	Host for tapeworm	Cysts in whitefish	Manitoba, Alberta	All year	Miller, 1950
Stoneroller	Competition on redds	Spawn of rainbow trout	Great Smoky Mountains National Park	Spring	Lennon and Parker, 1960
Goldfish	Competition	Trout	California	All year	Vestal, 1942
Carp	Competition, destroys habitat	Game fish, waterfowl, water quality	United States, Canada, Victoria, Australia	All year	Butcher, 1967; Sigler, 1958
Hitch	Competition	Trout, bass	California	All year	Burns, 1966

Sacramento squawfish	Predation	Pacific salmon	United States, Canada	Spring, summer	Taft and Murphy, 1950
Northern squawfish	Predation	Pacific salmon	United States, Canada	Spring, summer	Foerster and Ricker, 1941
White sucker	Competition	Game fish	United States	All year	LesVeaux, 1959
Longnose sucker	Competition	Game fish	Canada	All year	Rawson and Elsey, 1950
Largescale sucker	Competition	Game fish	United States, Canada	All year	Carl, 1936
Catfish spp.	Clog water lines	Economic	Texas, Missouri	All year	Gudger, 1950
Walking catfish	Predation	Game fish	Florida	All year	Idyll, 1969
Black bullhead	Competition	Game fish	Southern U.S.	All year	Viosca, 1931
American eel	Competition	Trout	Eastern Canada	All year	Smith and Saunders, 1955
	Clog water lines	Economic	New York, Nova Scotia, Massachusetts	Autumn	Gudger, 1950
Japanese eel	Predation	Cultured shrimp	Taiwan	All year	Tang, 1961
Mosquitofish	Predation on spawn	Cultured carp	India	Summer	Sreenivasan and Natarajan, 1962
Green sunfish	Predation	Game fish	Southern, western U.S.	All year	McKechnie and Tharratt, 1966
Pumpkinseed	Competition	Game fish	Eastern, midwest U.S.	All year	Jackson, 1956
Bluegill	Stunting	Game fish	California	All year	Emig, 1966
Largemouth bass	Exotic	Native fish	South America, East Africa	All year	Myers, 1947
Yellow perch	Predation	Game fish	Michigan, Wyoming	All year	Eschmeyer, 1936
	Competition	Game fish	Wisconsin	All year	Radonski, 1967
	Stunting	Game fish	New Hampshire	All year	Riel, 1965
Freshwater drum	Competition	Game fish	Wisconsin	All year	Priegel, 1967
Coastrange sculpin	Predation	Pink salmon	Alaska	Summer	Meehan and Sheridan, 1966
Torrent sculpin	Predation	Coho salmon	Washington	Summer	Patten, 1962

years, Chicago has experienced water shortages because of vast schools of alewives blocking the municipal water intake lines in Lake Michigan. The city is experimenting with compressed air curtains and blocking nets to keep the fish away from intakes. Mass mortalities of alewives in the spring and summer of 1967 gravely polluted the Canadian shore of Lake Erie and the shores of Lake Michigan.

Fishery managers hope that a general reduction of the alewife nuisance can be achieved by heavy commercial exploitation for meal and animal feed and by predation through stocking of Pacific salmons, rainbow trout, brook trout, and lake trout.

Carp

The successful introduction of carp into the United States from Germany less than a hundred years ago was hailed as a triumph (Laycock, 1966). Its spread throughout the country to provide food and sport was enthusiastically acclaimed, yet within a few years the popularity of the wonder fish waned rapidly. By the turn of the century it was recognized that carp was well and widely established and undoubtedly here to stay.

The carp is now considered to be the principal pest fish in the United States. It occurs in the 48 contiguous states, and is probably more abundant than any other freshwater fish. It is tough, wary, and exceedingly adaptable; it grows fast and large; and it constantly expands its range. The impact of this rough fish on the fishing economy is alarming (Sigler, 1958). Because of its abundance, large size, and rooting feeding habit, it causes constant and troublesome turbidities in once-clear lakes and rivers, wrecks aquatic vegetation important to waterfowl as habitat and food, and destroys fish spawning beds. For these and other reasons, it is the primary target of control efforts. A leading objective of the New York State Fish Laboratory at Livingston Manor and the federal Fish Control Laboratories at La Crosse, Wisconsin, and Warm Springs, Georgia, is the development of a reliable, safe, and preferably selective control for carp.

The many efforts to control populations of carp have included prohibition of stocking, the manipulation of water levels in impoundments, the erection of dams and barriers, seining, and use of toxicants. Successes are few. At the same time, the demand for carp has declined, with the result that commercial exploitation is falling off.

Goldfish

The problems with wild populations of the exotic goldfish are growing in the United States. Although smaller than carp, goldfish cause similar problems. Moreover, the species is usually more resistant than carp to fish toxicants. Its spread has been due to release of pet fish and to marketing of natural colored fish as bait minnows. Laws to prohibit such releases and marketing are in force in some states.

Largemouth Bass

The introduction of this important North American game fish into foreign waters has caused some serious problems. The predaceous bass has reduced valued, native fishes in South America. In Cuba it preyed heavily on native cyprinodont fishes that are important in malaria control (Rivero, 1936). It also damaged the important freshwater crab fishery in Lake Atitlan, Guatemala, and then extended its depredations to the young of the rare and flightless giant pied-billed grebe (*Podilymbus gigas*) (Powers and Bowes, 1967).

Other Fish

The expansion of pond-fish culture throughout the world engenders many problems with competing, predaceous, or disease-carrying fish. For example, raising channel catfish in artificial ponds for food and sport is a rapidly growing industry in southern United States, with a production of about 30 million pounds in 1967. In this monospecies culture, fish farmers desire to rid the ponds of invading minnows, sunfishes, and other fish because they consume significant quantities of food rations, carry diseases, and cause costly hand-sorting of fish at harvest time.

In Taiwan, several predaceous fishes require control in shrimp-culture ponds. The catfish, *Clarias* sp., is a troublesome predator in fish ponds in Malaya. The snakehead, *Ophicephalus* sp., requires some control in fish-culture ponds in Japan. Toxicants, barriers, nets, and water manipulation are among the controls employed with varying success. There is much room, however, for the development of safe and effective controls, and the great need for research on control measures is widely recognized.

Indeed, fishery literature provides abundant evidence that in many instances fish are serious pests. The foregoing examples provide only a sketchy résumé of important examples. Despite the fact that most of the problems with fish have been long-standing, man has done surprisingly little about them. This is primarily due to a lack of practical, safe, and economical means for correcting situations in which fish are pests. Also, both sport and commercial fishing have been pursued mostly by individuals, families, or small groups, and there has been little tendency among these typically independent people to combine and coordinate efforts to combat pest problems.

We are witnessing an increasing, often competing, use of waters for sport, recreation, agriculture, industry, and waste disposal; loss of fishing grounds to other uses; a pressing need to obtain greater yields of food and industrial fish

from the seas; a tremendous increase in farm fish ponds; and the recent strik-
ing growth of fishery research throughout the world. All these factors contribute to a better definition of problems having to do with fish and provide motivation and support toward their solution.

COMBATING PEST PROBLEMS

The approaches to fish control may be classed as biological, physical, chemical, and legal. Table 8 lists the controls applied to fishes in various pest situations and indicates the degree of success obtained.

BIOLOGICAL CONTROLS

Some of the earliest controls for undesirable fishes were biological controls.
Large predatory fishes have long been stocked to reduce stunted or unwanted
species, but with relatively few unqualified successes. In contrast, northern
pike and river perch are considered to be problem predators on desirable fish
in certain waters in Poland, and selective fishing for them is a means of control.

Attempts have been made to use infertile hybrids in cultivated fish ponds
to alleviate problems of overpopulation and stunting. C. F. Hickling at the
Tropical Fish Culture Research Institute, Malacca, Malaya, hybridized the
Mozambique tilapia with a different species from East Africa to produce all-
male offspring that have hybrid vigor and excellent growth (Tubb, 1967). This
monosex culture is receiving great attention in Uganda as a solution to over-
breeding of tilapia.

It is hoped that investigations in progress may disclose more biological controls for fish or integrated controls that include biological elements.

PHYSICAL CONTROLS

Mechanical

Man has unwittingly controlled or eradicated some of his prized fishes by
erecting dams. The Atlantic salmon, for example, was eliminated in colonial
New England by mill and logging dams on spawning streams. On the other
hand, dams, weirs, and diversions are employed to deny access of possible
pests into pest-free waters. The first and interim controls for sea lampreys in
the Great Lakes were temporary weirs that were installed in spawning streams.
Fish ladders can be designed and operated to permit ascent by desired fish but

prevent upstream movement by unwanted fish. Successful use of these measures, however, depends on favorable water conditions. Droughts, floods, ice, and debris may confound operations at critical times.

Seines, trap nets, and gill nets are employed with varying success in selected situations to catch and destroy problem fish. The State of Wisconsin, for example, issues large contracts for seining and destruction of carp. Explosives have been tried, but they yielded only limited success against sharks and gars. Too often the fish respond to these partial controls with vigorous, healthy replenishment that prolongs or even enlarges an aggravating problem.

The manipulation of water levels in streams and impoundments is a means of influencing fish populations that is coming into greater use wherever biologists can schedule the critical variations in levels. Water can be lowered immediately after spawning to leave the eggs and nests of certain undesirable fish high and dry. Also, water can be lowered at certain seasons to facilitate predation by piscivores on overabundant, forage species. Or, water levels may be stabilized or raised to aid desired species in spawning migrations or reproduction, thus promoting their dominance over unwanted fish.

Some of the newer ideas in mechanical controls include air curtains that are formed by release of compressed air from a submerged perforated tube or hose to repel or drive fish; light or sound for attracting or repelling; and baits for attracting and killing fish. Research on mechanical controls will continue to be stimulated by the necessity to keep masses of fish out of irrigation ditches, hydroelectric turbines, and water-intake lines, and from reinfesting reclaimed waters.

Electrical

Within the past two decades, electricity in alternating and direct currents has come into widespread use for collecting fish in marine and fresh waters (Applegate, Macy, and Harris, 1954). Most of the gear is fashioned to cause electronarcosis in fish, from which they may recover as soon as the electric field is removed.

The greatest use of electricity for control of fish has been in the sea lamprey program on the Great Lakes (Applegate, Smith, and Nielsen, 1952; McLain and Nielsen, 1953). Electrobarriers were installed in some streams to prevent upstream spawning migration by adult lampreys. Electromechanical weirs with traps were employed on other streams where commercial or game fish migrate concurrently with sea lampreys. This type of weir consists of an insulated fish trap located at one side or in the center of a stream; a fish- and lamprey-proof electrical field of alternating current extends laterally through the water, at right angles to stream flow, from the trap to the banks; and an electrical field of direct current extends diagonally downstream from the trap, with the posi-

TABLE 8 Controls Applied to Fishes in Pest Situations

Species	Control Program	Success[a]	Reference
Sea lamprey	Mechanical weirs	+	Applegate, 1950
	Electromechanical weirs	+++	Applegate, Smith, and Nielson, 1952
	Barrier dams	+++	Stauffer, 1964
	Downstream traps	++++	McLain and Manion, 1967
	Selective toxicants	++++	Applegate *et al.*, 1961
Man-eating sharks	Explosives	−	Ommanney, 1963
	Shark fences	+	Caras, 1964
	Patrols, watches	++	Caras, 1964
	Repellents	+	Caras, 1964
Spiny dogfish	Increased exploitation	−	Jensen, 1966
	Explosives	−	Jensen, 1966
Gars	Seining	+	Caldwell, 1912
	Explosives	+	Johnston, 1961
	Electrofishing	+	Burr, 1931
	Toxicants	+++	Gilderhus, Berger, and Lennon, 1969
Indian tarpon	Toxicants	+++	Tang, 1961
	Water level manipulation	+	Tang, 1961
Alewife	Increased exploitation	+	Moffett, 1966
	Stock predator fish	++	Stephenson, 1968
	Curtain of air bubbles	++	Kupfer and Gordon, 1966
	Barrier nets and screens	++	Gudger, 1950
Gizzard shad	Toxicants	+++	Jenkins, 1957
Dolly Varden trout	Bounties	−	Bower, 1943
	Weir-trapping	++	Rounsefell, 1958
	Gillnetting	+++	Foerster and Ricker, 1941
Piranha	Warning service	++	Caras, 1964
	Toxicants	+++	Fontenele, 1963

Carp	Commercial exploitation	−	Laycock, 1966
	Electrofishing	+	Burr, 1931
	Water level manipulation	++	Sigler, 1958
	Netting	+++	Ricker and Gottschalk, 1940
	Toxicants	++++	Gilderhus, Berger, and Lennon, 1969
Squawfishes	Explosives	+	Jeppson, 1957
	Toxicants	+	LesVeaux, 1959
	Gillnetting	+++	Foerster and Ricker, 1941
Suckers	Toxicants	+++	Foye, 1964
Bullheads	Toxicants	+++	Henegar, 1966
	Water level manipulation	+++	Viosca, 1931
Catfish (*Clarias* sp.)	Toxicants	+++	Sreenivasan and Natarajan, 1962
Candiru	Protective sheaths	++	Norman, 1963
American eel	Barrier weirs	+	Smith, 1955
	Baited traps	+	Smith, 1955
	Screens on intake lines	+	Gudger, 1950
Sunfishes	Toxicants	+++	Burress and Luhning, 1969
	Water level manipulation	++	Panikkar, 1960
Largemouth bass	Toxicants	+++	Powers and Bowes, 1967
Yellow perch	Toxicants	++++	Radonski, 1967
Snakeheads	Toxicants	+++	Sreenivasan and Natarajan, 1962
Striped mullet	Toxicants	+++	Tang, 1961
	Water level manipulation	++	Tang, 1961
Puffers	Regulations, inspection	+++	Walford, 1958
Parasitized freshwater food and game fishes	Cook fish thoroughly, prevent raw sewage in water, poison infected snails and fish	++++	Bauer, 1961

[a]Success of control is more often estimated than assessed; ++++ indicates highly successful, ranging to − (unsuccessful).

tive pole at the mouth of the trap and the negative pole across and downstream. The electrodes are mounted or suspended in such a manner that varying water levels and the passage of ice and debris past the weir do not interrupt the electrical fields. Fish and lampreys moving upstream first encounter the direct-current field and are led or deflected toward the positive electrode and into the trap. Those evading the direct current move on and encounter the impenetrable field of alternating current that deflects them to the trap, or forces them either to retire downstream or to remain within the field and perish. This electrical control of adult lampreys in conjunction with the chemical control of larval lampreys (to be discussed later) has been very effective.

In years to come, electricity undoubtedly will be used to a greater extent in fisheries. Fears of costs and of possible hazards somewhat deter developments and applications at present. Actually, in many circumstances the voltages and currents required to affect fish are modest and economical, and the hazards to other animals and humans are not insurmountable problems. Electric fences for fish might become as practical as electric fences for livestock.

CHEMICAL CONTROLS

Man long ago learned that certain plant toxins stun or kill marine and fresh-water fish. For more than 1,000 years, primitive peoples in Asia and the Americas have used rotenone-bearing roots of derris, cube, or timbo to collect food fishes. Similarly, saponin-bearing "fishing plants" have long been employed in Asia to collect fish in rivers and marine estuaries (Tang, 1961). It is also traditional among Chinese fish farmers to use tea-seed cake to eradicate undesirable fish from ponds before restocking. The cake is the residue remaining after oil is expressed from the seeds of *Camellia*, and it contains 10 to 13 percent of saponin. Moreover, Indians in the Americas applied crushed hulls of black walnut to streams to stun fish so that they might be collected.

Copper sulfate was the first toxicant used in sport fishery management in the United States (Titcomb, 1914). It is still used to a limited extent for control of cyprinids and other problem fish. It acts as an irritant to fish when applied in sublethal concentrations and can be employed in small lakes to drive fish so as to increase catches in fyke nets. The copper ion is also toxic to aquatic invertebrates and aquatic plants, and in soft waters it may be very persistent and cumulative.

Rotenone came into use as a fish toxicant in the 1930's, and the eradication of stunted or undesirable fish in lakes and streams began to be a common practice. The powdered derris root used at first was difficult to administer in water and its rotenone content varied, but fishery managers soon had emulsifiable, liquid, and synergized formulations of rotenone made available to them. Within the past two decades, many ponds, lakes, and streams in the United

States and Canada have been reclaimed with rotenone, indicating that rehabilitation of waters by poisoning fish may be the best available management tool (Prévost, 1960). Partial and spot poisonings of lakes are effective in certain circumstances to reduce fish problems.

Calcium hypochlorite has been used since 1947 in eliminating problem fish from water supply sources and small ponds. The ease of neutralizing it with sodium thiosulfate is considered an advantage.

Toxaphene, a chlorinated hydrocarbon insecticide, began to compete with rotenone as a fish toxicant in 1948. Unfortunately, it came into wide use in fish management before its disadvantages were apparent. Its hazards to waterfowl and other life and its long persistence in some waters caused the U.S. Bureau of Sport Fisheries and Wildlife and several states to discontinue its use after 1963 (Dykstra and Lennon, 1966). Furthermore, toxaphene is not registered by pesticide authorities as a fish toxicant, and it is unlikely that it could be.

Endrin, another insecticidal hydrocarbon compound, was tested on fish in 1952 and first applied as a general toxicant in 1958. After its use in eradicating fish in ponds in Asia, it was considered to be the most powerful fish poison known (Sreenivasan and Natarajan, 1962). A recent report states that endrin is applied in India to kill predatory and weed fishes in small fish-nursery ponds, after which the water is detoxified with relatively large quantities of charcoal (Bhimachar and Tripathi, 1967). The chemical, however, is not registered in the United States as a toxicant for fish and is not satisfactory for general use in fish control because it is highly toxic to other vertebrates and persists for a long time in water.

In 1956, fishery investigators in New Hampshire demonstrated that pellets of sodium hydroxide dropped in the nests of sunfish would kill eggs and fry (Jackson, 1956). Two years later, the efficacy of sodium cyanide as a fish poison was reported (Bridges, 1958). Although very hazardous to humans, the cyanide is toxic to fish at low concentrations, persists only briefly in water, and first anesthetizes fish before killing them. Exposed fish may survive if removed soon to fresh water.

The development of TFM (see Appendix B) as a selective lampricide is notable in two extremely important respects (Applegate *et al.*, 1961). It is the first chemical developed specifically as a fish control (others, such as rotenone, toxaphene, and endrin, are principally insecticides; their use as piscicides is secondary and relatively minor). Secondly, the use of TFM by the United States and Canada in streams tributary to the Great Lakes is a long step forward toward the control of the sea lamprey and the restoration of very important food and sport fisheries. Moreover, the success of TFM has stimulated investigations to find specific controls for other fish.

A malathion-dibrom mixture was introduced as a selective toxicant for certain sunfishes in 1962, but its usefulness appeared to be limited to relatively soft waters (Grice, 1962). At about the same time, studies were initiated to develop sperm toxins for the control of squawfish in west coast salmon streams (MacPhee, 1963).

Another compound that has been studied and developed for fish control is antimycin, an antibiotic produced by *Streptomyces*. It was approved and registered as a fish toxicant in Canada and the United States in 1966. Like TFM, antimycin has been studied thoroughly in the laboratory and field and found to be nonharmful to other aquatic life, waterfowl, or mammals at fish-killing concentrations (Herr, Greselin, and Chappel, 1967; Gilderhus, Berger, and Lennon, 1969). Antimycin is absorbed into the gills of fish, which it kills by interfering with the respiration of cells. It is effective in very small concentrations against fertilized eggs, fry, fingerling, and adult fish (Berger, Lennon, and Hogan, 1969). It imparts no color or odor to the water, and does not repel fish from treated areas as other toxicants tend to do. Since antimycin degrades rapidly in water, fish can be restocked relatively soon. Moreover, it appears to have possibilities for selective control of certain problem species without harming other fish in the same water (Burress and Luhning, 1969; Radonski, 1967).

Within recent years, Bayluscide (Bayer 73) has shown promise as a toxicant to snails that are vectors for schistosomiasis. The compound has been tested also as a toxicant for larval lampreys and fish in general (Marking and Hogan, 1967).

LEGAL CONTROLS

Legislation has not been used to any great extent in the alleviation of pest-fish situations. Often a local but serious problem occurs in international, interstate, or public waters over which the community concerned has no valid jurisdiction, and often a pest situation worsens—or dissipates—before the lengthy process of securing legislation to enable control can be initiated.

The control of the sea lamprey in the Great Lakes is a good example of legal processes needed to bring about required action and cooperation. The commercial fisheries and supporting industries of many small communities on the upper Great Lakes were hurt badly by the depredations of the lamprey, but there was little they could do except bring pressure to bear on their legislators. The states and provinces had no appropriate pacts under which they could make a joint attack on the very large problem, and they "passed the buck" to their federal governments. The federal authorities proceeded to order biological investigations of the sea lamprey and, at the same time, to explore possibilities for an international agreement regarding its control. After nearly a decade of negotiations among state, provincial, and federal officials, the United States and Canada formed the Great Lakes Fishery Commission to combat the sea lamprey, to restore populations of food and game fish, and to manage the fishery resources of the lakes. The legislative process was long and difficult, and it is fortunate indeed that the research on control measures and early applications of electrical and chemical controls had moved ahead in the meantime under the auspices of the U.S. and Canadian fishery agencies.

In 1928 the salmon interests in Alaska secured legislation to permit paying bounties on the predaceous Dolly Varden trout, but the practice was abandoned in 1941 (Bowar, 1943). The payment of bounties on the spiny dogfish, often a serious pest in the Gulf of Maine, has been suggested but has not been adopted (Jensen, 1966).

Within recent years, there has been a growing tendency to modify existing laws governing fisheries to permit greater sport and commercial utilization of rough and potentially pest fish. For example, archery fishing for carp and spearing of spawning-migrant suckers are on the increase as sports. Laws are relaxed at times in eastern Europe to foster larger harvests of overly-abundant predaceous fish.

One of the most important areas for legal action is the regulation of importing and stocking live fish. Tighter controls may be warranted to restrict the introduction of exotic fishes (Gottschalk, 1967). There is risk that imports of tropical fishes for the flourishing home-aquarist trade will lead to the escape or release of dangerous species, such as piranha, into public waters (Laycock, 1966). The walking catfish is a recent example of an unfortunate introduction of an exotic into Florida (Idyll, 1969). The interstate and even international spread of fish diseases, especially among domesticated trouts, has provoked some demand for regulations to require inspection, disinfection, or even prohibition of imported fish. The use of certain fish as bait, such as carp and goldfish, is prohibited in some states. In many waters that have been reclaimed to correct a pest situation, the use of any live fish as bait is illegal. Moreover, it is commonly illegal for anyone but authorized individuals to stock fish in public waters.

In every case, the laws and regulations affecting the control of fish should be based on sound biological information. Too often they have run counter to biological principles, and have been ineffective, unrealistic, or unenforceable.

METHODOLOGY AND MECHANICS

Effective animal control is the translation of ecology into policy (Howard, 1967). The control of problem fish or of situations in which fish are pests requires the following:

    ● Adequate knowledge of the life history of each target species in order that control efforts may be selected and directed to the more vulnerable stage of life.

    ● Adequate knowledge of the ecological, human health, safety, agricultural, industrial, recreational, urban, or other factors in situations where fish are undesirable, so that correction of such a factor may reduce or eliminate the problem.

● An arsenal of specific controls—biological, physical, and chemical—that can be used singly or in combination for selective or general correction of fish problems in various marine and freshwater situations.

● Efficient and economical means for applying controls and assessing the effects.

At present, there are serious gaps in each of these requirements. Until they are filled, many efforts in fish control will remain relatively unsophisticated. Fortunately, recent developments in the fields of fisheries, oceanography, limnology, water pollution, and electronics have contributed a battery of new aids for studying fish and their habitats. Electrofishing apparatus is used to sample fish populations; techniques and portable instrumentation are available for on-site chemical analyses of water; field-type electric thermometers yield almost instant information on temperatures at depths and on thermal stratification of water; new portable fathometers greatly facilitate the essential determinations of underwater topography; and minisubmarines, scuba gear, and underwater television permit observations heretofore impossible.

Several fishery research laboratories are now exclusively engaged in finding and developing controls specific to fish (Lennon and Walker, 1964). In addition, other laboratories are becoming involved in fish control. Most of these investigations originated within the past 15 years and have resulted, for example, in the development of electroweirs, improved fish ladders, TFM as a lampricide, and antimycin as a fish toxicant. Developments in the offing include vigorous and more competitive hybrid fish for commercial and sport fish culture; selective breeding in fish culture to improve survivability of fish stocked in the wild; selected uses of predaceous exotics to control problem forage fish and increase commercial and sport fishing opportunity; additional and possibly selective fish toxicants; and improved formulations of fish toxicants to enhance their applicability and effectiveness in a wide variety of situations.

Also, significant improvements in the mechanics of fish control have been made in recent years. Lighter or better metals are used in weirs and electroweirs; power sources for electrogear have been made more portable and longer lasting; synthetic fibers are proving more effective and longer lasting in nets and seines; portable apparatus is available for underwater aeration and air curtains; and helicopters and fixed-wing aircraft are increasingly employed in large-scale applications of toxicants.

Economics are extremely important in animal control, and it appears that costs are a greater deterrent to control of fish than of other animals. Fishermen, water sports enthusiasts, and boaters use public waters to a large extent, and there are few strong organizations among them that can surmount inter-community, interstate, or international politics to press for correction of fish problems. To the public at large, fish problems are underwater and out of sight; therefore, small groups or agencies have difficulty arousing support for corrective measures until a problem has become very grave. As a result, many attempts at fish control have been compromised to an inadequate approach with poor success because of the necessity to keep costs down to an imprac-tical minimum.

Also, a new cost burden was added to chemical control agents several years ago when the congress of the United States strengthened regulations concern-ing all pesticides. Any chemical used in fish control must now be approved and registered for the purpose (Lennon, 1967). This action is justifiable from every point of view; however, the research required to meet the higher standards for approval is both lengthy and expensive, and it raises the costs of products. Among the chemicals used as fish toxicants, rotenone, TFM, and antimycin are registered; toxaphene, endrin, copper sulfate, calcium hypochlorite, and sodium cyanide are not.

The relatively high costs of approved and registered fish toxicants may not be such a great disadvantage if they force the users to make more complete appraisals of pest problems, to better define target situations, to seek biologi-cal or integrated controls, or to find less stringent alternatives. The price of toxicant for a lake or stream reclamation is only a fraction of the total cost of the operation, but if the chemical is cheap there is greater likelihood of abuse and overdose.

APPENDIX A: COMMON AND SCIENTIFIC NAMES OF FISHES

Common Name	Scientific Name	Common Name	Scientific Name
Barracudas	Sphyraenidae	Hagfishes	Myxinidae
Great barracuda	*Sphyraena barracuda*	Herrings	Clupeidae
Bowfins	Amiidae	Alewife	*Alosa pseudoharengus*
Bowfin	*Amia calva*	Gizzard shad	*Dorosoma cepedianum*
Catfishes		Jacks	Carangidae
	Ictaluridae	Lampreys	Petromyzonidae
Black bullhead	*Ictalurus melas*	Sea lamprey	*Petromyzon marinus*
	Pygidiidae	Livebearers	Poeciliidae
Candirus (Carnero)	*Vandellia* spp.	Mosquitofish	*Gambusia affinis*
	Clariidae	Mackerels and tunas	Scombridae
Clarias	*Clarias* spp.	Bonitos	*Sarda* spp.
Walking catfish	*Clarias batrachus*	Skipjacks	*Euthynnus* spp.
Characins	Characidae	Tunas	*Thunnus* spp.
Piranhas (Caribe)	*Serrasalmus* spp.	Minnows and carp	Cyprinidae
	Rooseveltiella sp.	Barbels	*Barbus* spp.
	Pygocentrus sp.	Bream	*Abramis brama*
Chimaeras (Rabbitfishes)	Chimaeridae	Carp	*Cyprinus carpio*
Cichlids	Cichlidae	European dace	*Leuciscus leuciscus*
Mozambique tilapia	*Tilapia mossambica*	Goldfish (silver crucian carp)	*Carassius auratus*
Eels (freshwater)	Anguillidae	Hitch	*Lavinia exilicauda*
American eel	*Anguilla rostrata*	Ide	*Leuciscus idus*
European eel	*Anguilla anguilla*	Northern squawfish	*Ptychocheilus oregonensis*
Japanese eel	*Anguilla japonica*	Roach	*Rutilus rutilus*
Electric eels	Electrophoridae	Rudd	*Scardinius erythrophthalmus*
Electric eel	*Electrophorous electricus*	Sacramento squawfish	*Ptychocheilus grandis*
Electric rays	Torpedinidae	Stoneroller	*Campostoma anomalum*
Drums	Sciaenidae	Tench	*Tinca tinca*
Freshwater drum	*Aplodinotus grunniens*	White amur	*Ctenopharyngodon idella*
Gars	Lepisosteidae		

34

Common name	Scientific name	Common name	Scientific name
Morays	Muraenidae	Spiny dogfish	*Squalas acanthias*
Green moray	*Gymnothorax funebris*	Snake eels	Opichthidae
Mullets	Mugilidae	Serpent eel	*Opichthys serpens*
Striped mullet	*Mugil cephalus*	Snakeheads	Ophicephalidae
Oarfish	Regalecidae	Snappers	Lutjanidae
Perches	Percidae	Stingrays	Dasyatidae
Common perch	*Perca fluviatilis*	Suckers	Catostomidae
Ruff	*Acerina cernua*	Largescale sucker	*Catostomus macrocheilus*
Walleye	*Stizostedion vitreum*	Longnose sucker	*Catostomus catostomus*
Yellow perch	*Perca flavescens*	White sucker	*Catostomus commersoni*
Pikes	Esocidae	Sunfishes	Centrarchidae
Chain pickerel	*Esox niger*	Bluegill	*Lepomis macrochirus*
Northern pike	*Esox lucius*	Green sunfish	*Lepomis cyanellus*
Puffers	Tetraodontidae	Largemouth bass	*Micropterus salmoides*
Remoras	Echeneidae	Pumpkinseed	*Lepomis gibbosus*
Scorpionfishes	Scorpaenidae	Surgeonfishes	Acanthuridae
Rockfishes	*Sebastodes* spp.	Tarpons	Elopidae
Scorpionfishes	*Scorpaena* spp.	Indian tarpon	*Megalops cyprinoides*
Stonefish	*Synanceja verru cosa*	Toadfishes	Batrachoididae
Sculpins	Cottidae	Trouts, whitefishes	Salmonidae
Coastrange sculpin	*Cottus aleuticus*	Brook trout	*Salvelinus fontinalis*
Torrent sculpin	*Cottus rhotheus*	Brown trout	*Salmo trutta*
Sea basses	Serranidae	Dolly Varden	*Salmo malma*
Sharks	Carchariidae	Lake trout	*Salvelinus namaycush*
	Carcharhinidae	Whitefishes	*Coregonus* spp.
	Lamnidae	Weeverfishes	Trachinidae
	Sphyrinidae	Great weever	*Trachinus draco*
	Squalidae	Wrasses	Labridae

Sources: American Fisheries Society. 1960. A list of common and scientific names of fishes from the United States and Canada. Special Publication No. 2, second edition, American Fisheries Society, Washington, D.C. 102 pp.

Herald, E. S. 1961. Living fishes of the world. Doubleday and Company, Garden City, New York, 304 pp.

Sterba, G. 1963. Freshwater fishes of the world. The Viking Press, New York. 878 pp. (Translated from German).

APPENDIX B: FORMULAS OF FISH TOXICANTS

Toxicant	Formula
Antimycin A[a]	$C_{28}H_{40}N_2O_9$
Bayluscide (Bayer 73)	2-aminoethanol salt of 2',5-dichloro-4'-nitrosalicylanilide
Calcium hypochlorite	$Ca(OCl)_2 \cdot 2H_2O$
Copper sulfate	$CuSO_4$
Endrin	1,2,3,4,10,10-hexachloro-6,7-epoxy-1,4,4a,5,6,7,8,8a-octahydro-1,4,5,8-*endo-endo*-dimethanonaphthalene
Malathion-Dibrom mixture	S-(1,2-dicarbethoxyethyl)-*O,O*-dimethyldithiophosphate and *O,O*-dimethyl-*O*-(1,2 dibromo-2,2-dichloroethyl) phosphate
Rotenone[a]	$C_{23}H_{22}O_6$
Saponins	Sapogenin glycosides, triterpenoid saponins (structure not known)
Sodium cyanide	NaCN
Sodium hydroxide	NaOH
T FM[a]	3-trifluormethyl-4-nitrophenol
Toxaphene	$C_{10}H_{10}Cl_8$

[a]Approved and registered for fishery use in the United States.

REFERENCES

Applegate, V. C. 1950. Natural history of the sea lamprey, *Petromyzon marinus*, in Michigan. U.S. Fish and Wildlife Service, Special Scientific Report–Fisheries No. 55. 237 pp.

Applegate, V. C., J. H. Howell, J. W. Moffett, B. G. H. Johnson, and M. A. Smith. 1961. Use of 3-trifluormethyl-4-nitrophenol as a selective sea lamprey larvicide. Great Lakes Fishery Comm., Technical Report No. 1. 35 pp.

Applegate, V. C., P. T. Macy, and V. E. Harris. 1954. Selected bibliography on the applications of electricity in fishery science. U.S. Fish and Wildlife Service, Special Scientific Report–Fisheries No. 127. 55 pp.

Applegate, V. C., B. R. Smith, and W. L. Nielsen. 1952. Use of electricity in the control of sea lampreys: electromechanical weirs and traps and electrical barriers. U.S. Fish and Wildlife Service, Special Scientific Report–Fisheries No. 92. 52 pp.

Armstrong, G. C. 1949. The removal of undesirable fish from the headwaters of Dorion Rearing Station in 1948. Canadian Fish Cult. 4(5):7–10.

Baldwin, N. S. 1968. Sea lamprey in the Great Lakes. Limnos. 1(3):20–27.

Bauer, O. N. 1961. Fishes as carriers of human helminthoses, pp. 320–334. *In* V. A. Dogiel, G. K. Petrushevski, and Yu. I. Polyanski (eds.). Parasitology of fishes. Oliver and Boyd, Ltd., Edinburgh and London. (Translated from Russian.)

Belding, D. L. 1942. Textbook of clinical parasitology. Appleton-Century-Crofts, New York. 888 pp.

Berger, B. L., R. E. Lennon, and J. W. Hogan. 1969. Investigations in fish control: 26. Laboratory studies on antimycin as a fish toxicant. U.S. Bureau of Sport Fisheries and Wildlife. 21 pp.

Bhimachar, B. S., and S. D. Tripathi. 1967. A review of culture fisheries activities in India, pp. 1–33. *In* T. V. R. Pillay (ed.). Proceedings of the FAO world symposium on warm-water pond fish culture. FAO Fisheries Report 2(44). (Regional and Country Report.)

Bigelow, H. B., and W. C. Schroeder. 1948. Sharks, pp. 59–546. *In* J. Tee-van (ed.). Fishes of the Western North Atlantic. Part I. Sears Foundation for Marine Research, Yale University, New Haven.

Bigelow, H. B., and W. C. Schroeder. 1953. Fishes of the Gulf of Maine. U.S. Department of the Interior, Fishery Bulletin of the Fish and Wildlife Service 53(74). 577 pp.

Bower, W. T. 1943. Alaska fishery and fur-seal industries: 1941. Statistical Digest No. 5, U.S. Fish and Wildlife Service, pp. 1–71.

Bridges, W. R. 1958. Sodium cyanide as a fish poison. U.S. Fish and Wildlife Service, Special Scientific Report–Fisheries No. 253. 11 pp.

Burns, J. W. 1966. Hitch, pp. 520–522. *In* A. Calhoun (ed.). Inland fisheries management. California Department of Fish and Game, Sacramento.

Burr, J. G. 1931. Electricity as a means of garfish and carp control. Trans. Amer. Fish. Soc. 31:174–183.

Burress, R. M., and C. W. Luhning. 1969. Investigations in fish control: 28. Use of antimycin for selective thinning of sunfish populations in ponds. U.S. Bureau of Sport Fisheries and Wildlife. 10 pp.

Butcher, A. D. 1967. A changing aquatic fauna in a changing environment, pp. 197–218. *In* International Union for Conservation of Nature and Natural Resources. Towards a new relationship of man and nature in temperate lands. Part III. Changes due to introduced species. Morges, Switzerland, New Series No. 9.

Caldwell, E. E. 1912. The gar problem. Trans. Amer. Fish. Soc. 42:61–65.

Caras, R. A. 1964. Dangerous to man. Chilton Books, Philadelphia and New York. 433 pp.

Carl, G. C. 1936. Food of the coarse-scaled sucker (*Catostomus macrocheilus* Girard). J. Biol. Board of Canada 3(1):20–25.

Davis, R. M. 1967. Parasitism by newly-transformed anadromous sea lampreys on land-locked salmon and other fishes in a coastal Maine lake. Trans. Amer. Fish. Soc. 96(1):11–16.

Dykstra, W. W., and R. E. Lennon. 1966. The role of chemicals for the control of verte-brate pests, pp. 29–34. *In* E. F. Knipling (chairman). Pest control by chemical, bio-logical, genetic, and physical means. A symposium. U.S. Department of Agriculture, ARS 33–110.

Emig, J. W. 1966. Bluegill sunfish, pp. 375–392. *In* A. Calhoun (ed.). Inland fisheries management. California Department of Fish and Game, Sacramento.

Eschmeyer, R. W. 1936. Some characteristics of a stunted perch population. Papers Michigan Acad. Sci., Arts, and Letters 22:613–628.

Foerster, R. E., and W. Ricker. 1941. The effect of reduction of predaceous fish on sur-vival of young sockeye salmon at Cultus Lake. J. Fish. Res. Board of Canada 5(4):315–336.

Fontenele, O. 1963. Eradication of piranha in inland waters. Com. Fish. Rev. 25(3):46–50.

Foye, R. E. 1964. Chemical reclamation of forty-eight ponds in Maine. Progressive Fish-Cult. 26(4):181–185.

Ghiretti, F., and E. Rocca. 1963. Some experiments on ichthyotoxin, pp. 211–216. *In* H. L. Keegan and W. V. Macfarlane (eds.). Venomous and poisonous animals and noxious plants of the Pacific region. The Macmillan Co., New York.

Gilderhus, P. A., B. L. Berger, and R. E. Lennon. 1969. Investigations in fish control: 27. Field trials of antimycin A as a fish toxicant. U.S. Bureau of Sport Fisheries and Wildlife. 21 pp.

Gottschalk, J. S. 1967. The introduction of exotic animals into the United States, pp. 124–140. *In* International Union for Conservation of Nature and Natural Resources. Towards a new relationship of man and nature in temperate lands. Part III. Changes due to introduced species. Morges, Switzerland, New Series No. 9.

Grice, F. 1962. Field tests of Ortho Fish Thinner in Massachusetts. Abstract, Paper pre-sented at Northeast Section, American Fisheries Society, Monticello, N.Y., May 16, 1962.

Grosvenor, M. B. 1965. Wondrous world of fishes. Natl Geographic Society, Washington, D.C. 367 pp.

Gudger, E. W. 1930. Bookshelf browsing on the alleged penetration of the human urethra by an Amazonian catfish candiru. Part I. Amer. J. of Surgery N.S. 8(1):170–188. Part II. N.S. 8(2):443–457.

Gudger, E. W. 1950. Fishes in water pipes. Accounts of troubles caused by these intruders and the why and how of this behavior. Amer. Midland Natur. 43:399–409.

Halstead, B. W. 1962. Biotoxications, allergies, and other disorders, pp. 521–540. *In* G. Borgstrom (ed.). Fish as food. Vol. 2. Academic Press, New York.

Halstead, B. W. 1965. Poisonous and venomous marine animals of the world. Vol. 1, Invertebrates. U.S. Government Printing Office, Washington, D.C. 994 pp.

Halstead, B. W., and L. R. Mitchell. 1963. A review of the venomous fishes of the Pacific Area, pp. 173–202. *In* H. L. Keegan and W. V. Macfarlane (eds.). Venomous and poisonous animals and noxious plants of the Pacific Region. The Macmillan Co., New York.

Henegar, D. L. 1966. Investigations in fish control: 3. Minimum lethal levels of toxaphene as a piscicide in North Dakota lakes. U.S. Bureau of Sport Fisheries and Wildlife, Re-source Publ. No. 7. 16 pp.

Herald, E. S. 1961. Living fishes of the world. Doubleday and Co., Garden City, New York. 304 pp.

Herr, F., E. Greselin, and C. Chappel. 1967. Toxicology studies of antimycin, a fish eradicant. Trans. Amer. Fish Soc. 96(3):320-326.

Hildebrand, S. F., and W. C. Schroeder. 1928. Fishes of Chesapeake Bay. U.S. Department of Commerce, Bulletin of the Bureau of Fisheries (1927), Part I. 43(1). 366 pp.

Howard, W. E. 1967. Vertebrate pests: Biocontrol and chemosterilants, pp. 343–386. *In* W. W. Kilgore and R. L. Doutt (eds.). Pest control: Biological, physical, and selected chemical methods. Academic Press, New York.

Idyll, C. P. 1969. New Florida resident, the walking catfish. Natl. Geographic 135(6):864 851.

Jackson, C. F. 1956. Control of the common sunfish or pumpkinseed, *Lepomis gibbosus*, in New Hampshire. New Hampshire Fish and Game Department, Concord, Tech. Circ. No. 12. 16 pp.

Jenkins, R. M. 1957. The effect of gizzard shad on the fish population of a small Oklahoma lake. Trans. Amer. Fish. Soc. (1955) 85:58–74.

Jensen, A. C. 1966. Life history of the spiny dogfish. U.S. Bureau of Commercial Fisheries, Fishery Bull. 65(3):527–554.

Jeppson, P. 1957. The control of squawfish by use of dynamite, spot treatment and reduction of lake levels. Progressive Fish-Cult. 19(4):168–171.

Johnston, K. H. 1961. Removal of longnose gar from rivers and streams with the use of dynamite. Proc. Fifteenth Ann. Conf. Southeastern Assoc. Game and Fish Comm., Atlanta, Georgia, October 22–25, 1961: 205–207.

Kupfer, G. A., and W. G. Gordon. 1966. An evaluation of the air bubble curtain as a barrier to alewives. Comm. Fish. Rev. 28(9):1–9.

Lagler, K. F., J. E. Bardach, and R. R. Miller. 1962. Ichthyology. John Wiley and Sons, New York. 545 pp.

Lagler, K. F., and A. T. Wright. 1962. Predation of the Dolly Varden, *Salvelinus malma*, on young salmons, *Oncorhynchus* spp., in an estuary of southeastern Alaska. Trans. Amer. Fish. Soc. 91(1):90–93.

La Gorce, J. O. 1952. The book of fishes. National Geographic Society, Washington, D.C. 339 pp.

Laycock, G. 1966. The alien animals. The Natural History Press, Garden City, New York. 240 pp.

Lennon, R. E. 1967. The clearance and registration of chemical tools for fisheries. Progressive Fish-Cult. 29(4):187–193.

Lennon, R. E., and P. S. Parker. 1960. The stoneroller, *Campostoma anomalum* (Rafinesque), in Great Smoky Mountains National Park. Trans. Amer. Fish Soc. 89(3):263-270.

Lennon, R. E., and C. R. Walker. 1964. Investigations in fish control: 1. Laboratories and methods for screening fish-control chemicals. U.S. Bureau of Sport Fisheries and Wildlife, Circ. 185. 15 pp.

LesVeaux, J. F. 1959. Summary report of survey to evaluate the need for specific fish toxicants in sport fishing waters. Progressive Fish-Cult. 21(3):99–110.

MacPhee, C. 1963. The determination and development of sperm toxins for the control of undesirable species of fish. U.S. Bureau of Commercial Fisheries, Columbia River Fishery Development Prog., Final Rep. for Fiscal Year 1963 (mimeo). 14 pp.

Mansueti, R. J. 1962. Distribution of small, newly metamorphosed sea lampreys, *Petromyzon marinus*, and their parasitism of menhaden, *Brevoortia tyrannus*, in mid-Chesapeake Bay during winter months. Chesapeake Sci. 3(2):137–139.

Marking, L. L., and J. W. Hogan. 1967. Investigations in fish control: 19. Toxicity of
   Bayer 73 to fish. U.S. Bureau of Sport Fisheries and Wildlife, Resource Publ. 36. 13 pp.
Marshall, N. B. 1966. The life of fishes. The World Publishing Co., Cleveland. 402 pp.
McKechnie, R. J., and R. C. Tharratt. 1966. Green sunfish, pp. 399–401. *In* Alex Calhoun
   (ed.). Inland fisheries management. California Department of Fish and Game, Sacra-
   mento.
McLain, A. L., and P. J. Manion. 1967. An all-season trap for downstream-migrating fish
   and other aquatic organisms. Progressive Fish-Cult. 29(2):114–117.
McLain, A. L., and W. L. Nielsen. 1953. Directing the movement of fish with electricity.
   U.S. Fish and Wildlife Service, Special Scientific Report—Fisheries No. 93. 24 pp.
Meehan, W. R., and W. L. Sheridan. 1966. Investigations in fish control: 8. Effects of
   toxaphene on fishes and bottom fauna of Big Kitoi Creek, Afognak Island, Alaska.
   U.S. Bureau of Sport Fisheries and Wildlife, Resource Publ. 12. 9 pp.
Miller, R. B. 1950. The Square Lake experiment: An attempt to control *Triaenophorus
   crassus* by poisoning pike. Canadian Fish Cult. 7:3–18.
Moffett, J. W. 1966. Fishery problems in the Great Lakes. Comm. Fish. Rev. 28(2):48–49.
Myers, G. S. 1947. Foreign introductions of North American fishes. Progressive Fish-Cult.
   9(4):177–180.
Nielson, R. S. 1953. Should we stock brown trout? Progressive Fish-Cult. 15(3):125–126.
Nikol'skii, G. V. 1961. Special ichthyology. National Science Foundation and the Smith-
   sonian Institution, Washington, D.C. 538 pp. (Translated from Russian.)
Norman, J. R. 1963. A history of fishes. Second edition by P. H. Greenwood. Hill and
   Wang, New York. 398 pp.
Ommanney, F. D. 1963. The fishes. Time Inc., New York. 192 pp.
Oppenheimer, C. H. 1962. On marine fish diseases, pp. 541–572. *In* G. Borgstrom (ed.).
   Fish as food. Vol. 2. Academic Press, New York.
Panikkar, B. M. 1960. Low concentrations of calcium hypochlorite as a fish and tadpole
   poison applicable for use in partly drained ponds and other small bodies of water.
   Progressive Fish-Cult. 22(3):117–120.
Patten, B. G. 1962. Cottid predation upon salmon fry in a Washington stream. Trans.
   Amer. Fish. Soc. 91(4):427–429.
Powers, J. E., and A. LaB. Bowes. 1967. Elimination of fish in the Giant Grebe Refuge,
   Lake Atitlan, Guatemala, using the fish toxicant, antimycin. Trans. Amer. Fish. Soc.
   96(2):210–213.
Prévost, G. 1960. Use of fish toxicants in the Province of Quebec. Canadian Fish Cult.
   28:13–35.
Priegel, G. R. 1967. Identification of young walleyes and saugers in Lake Winnebago,
   Wisconsin. Progressive Fish-Cult. 29(2):108–109.
Radonski, G. C. 1967. Antimycin: Useful in perch control? Wisconsin Conserv. Bull.
   32(2):15–16.
Randolph, M. J. 1965. Roughnecks three. Virginia Wildlife 16(5):6,7,22.
Rawson, D. C., and C. A. Elsey. 1950. Reduction in the longnose sucker population of
   Pyramid Lake, Alberta, in an attempt to improve angling. Trans. Amer. Fish. Soc.
   (1948) 78:13–31.
Ricker, W. E., and J. Gottschalk. 1940. An experiment in removing coarse fish from a
   lake. Trans. Amer. Fish. Soc. 70:382–391.
Riel, A. D. 1965. The control of an overpopulation of yellow perch in Bow Lake, Straf-
   ford, New Hampshire. Progressive Fish-Cult. 27(1):37–41.
Rivero, L. H. 1936. The introduced largemouth bass, a predator upon native Cuban fishes.
   Trans. Amer. Fish. Soc. 66:367–368.

Rounsefell, G. A. 1958. Factors causing decline in sockeye salmon of Karluk River, Alaska. U.S. Fish and Wildlife Service, Fishery Bull. 58(130):83–169.

Schultz, L. P. 1967. Predation of sharks on man. Chesapeake Sci. 8(1):52–56.

Sigler, W. F. 1958. The ecology and use of carp in Utah. Utah State Univ., Agri. Exp. Sta., Logan, Bull. 405. 63 pp.

Smith, M. W. 1950. The use of poisons to control undesirable fish in Canadian fresh waters. Canadian Fish Cult. 8:17–29.

Smith, M. W. 1955. Fertilization and predator control to improve trout angling in natural lakes. J. Fish. Res. Board of Canada 12(2):211–237.

Smith, M. W., and J. W. Saunders. 1955. The American eel in certain fresh waters of the maritime provinces of Canada. J. Fish. Res. Board of Canada 12(2):238–269.

Soljan, T. 1963. Fishes of the Adriatic. Nolit Publishing House, Belgrade, Yugoslavia. 428 pp. (Translated from Serbo-Croatian).

Solman, V. E. F. 1945. The ecological relations of pike, *Esox lucius* L., and waterfowl. Ecology 26(2):157–170.

Sreenivasan, A., and M. V. Natarajan. 1962. Use of endrin in fishery management. Progressive Fish-Cult. 24(4):181.

Stauffer, T. M. 1964. An experimental sea lamprey barrier. Progressive Fish-Cult. 26(2):80–83.

Stephenson, W. J. 1968. Coho. Coho Unlimited, Kalamazoo, Michigan. 64 pp.

Taft, A. C., and G. I. Murphy. 1950. The life history of the Sacramento squawfish (*Ptychocheilus grandis*). California Fish and Game 36(2):147–164.

Tang, Y. 1961. The use of saponin to control predaceous fishes in shrimp ponds. Progressive Fish-Cult. 23(1):43–45.

Thompson, J. R., and S. Springer. 1961. Sharks, skates, rays, and chimaeras. U.S. Bureau of Commercial Fisheries, Circ. 119. 19 pp.

Titcomb, J. W. 1914. The use of copper sulfate for the destruction of obnoxious fishes in ponds and lakes. Trans. Amer. Fish. Soc. 44:20–24.

Tubb, J. A. 1967. Status of fish culture in Asia and the Far East. pp. 45–53. *In* T. V. R. Pillay (ed.). Proc. of the FAO world symposium on warm-water pond fish culture. FAO Fisheries Rep. 2(44) (Regional and Country Rep.).

U.S. Bureau of Commercial Fisheries. 1955. Ram on vessel's prow kills sharks. Comm. Fish. Rev. 17(11):48–49.

Van Duijn, C., Jr. 1962. Diseases of fresh-water fish. pp. 573–593. *In* G. Borgstrom (ed.). Fish as food. Vol. 2. Academic Press, New York.

Vestal, E. H. 1942. Reclamation with rotenone of Crystal Lake, Los Angeles County, California. California Fish and Game 28(3):136–142.

Viosca, P., Jr. 1931. The bullhead, *Ameiurus melas catulus*, as a dominant in small ponds. Copeia 1:17–19.

Walford, L. A. 1958. Living resources of the sea. The Ronald Press, New York. 321 pp.

CHAPTER 3

Amphibians and Reptiles as Pests

GEOGRAPHICAL DISTRIBUTION

Amphibians and reptiles are essentially tropical and subtropical. Darlington (1957) is an excellent source on their biogeography. Numbers of indigenous species dwindle rapidly and all but vanish as one moves poleward through temperate latitudes. Both groups are about equally unable to penetrate cooler climates. In the north, very few species of each group reach or pass the Arctic Circle, but above latitude 50°N there is only a very rudimentary fauna in either. To the south, both groups reach the southernmost extensions of continental land.

The amphibians are essentially continental animals, showing no ability to invade truly marine habitats, and only the feeblest ability to disperse across any extensive reaches of ocean. Reptiles also are primarily continental, although a few species, well distributed among the major orders, are strictly marine, and many others are characteristic of environments along the continental edges—on shores, in coastal marshes, and in estuaries. Reptiles are more often members of insular faunas than amphibians. Although little is known of the means employed by widely distributed reptiles in crossing salt-water barriers, it is clear that they are more tolerant of conditions encountered in such crossings.

The distribution of amphibians on the continents appears to be correlated with the distribution of water. Most amphibians, terrestrial as adults, require free water in which to pass egg and larval stages. More humid parts of the continents, where the temperature is suitable, tend to have amphibian faunas richer in species than drier ones. A number of species, however, through special adaptations of the egg and larval stages, are able to minimize or even eliminate the free-water requirement (e.g., *Plethodon cinereus*, the red-backed salamander of eastern United States). In some extremely dry environments (North American short-grass steppe or "desert"), a few specialized species of amphibians may be represented by dense populations, a fact often unsuspected by anyone who has not been present immediately after a rare shower or who has made no night observations. In contrast, reptiles are admirably adapted to occupy xeric habitats and are commonly represented in dry regions by larger assemblages of species than amphibians. Like the amphibians, however, they produce increasing numbers of species in more humid environments and so have a generally similar distribution in relation to the availability of water.

The relation between richness of amphibian and reptile faunas and density of human population is complex. In the New World and in African tropics, where densities of human beings are relatively low, there are many species of amphibians and reptiles. The densely populated North American, European, and Asiatic temperate zones have relatively few amphibian and reptile species. In the subtropics and tropics of southeast Asia and in tropical India and Ceylon, high human population densities and rich amphibian and reptilian faunas occur together. Thus, in some parts of the world, human populations are in contact with numerous species of amphibians and reptiles, and have been for a long time; elsewhere, large numbers of human beings have been in contact with only a small number of different species; and in some very few places, humans have been in contact with none at all. The strongest injurious effects of venomous snakes on human beings occur in the tropics and subtropics of India, Ceylon, and southeast Asia, where dense human populations and rich reptilian faunas occur together.

AMPHIBIANS

Amphibians are singular in the tenuousness of their association with man. As Cochran (1961) notes, they

. . . are the least harmful to man of any major zoological class. None of them has a poisonous bite, as do some of the snakes; they do not rob fields of grain and fruit, as do many birds; none of them is ferocious, like the carnivores, or destructive, like some rodents; they do not have such poisoned barbs as some fish possess, and they have inflicted no major diseases upon us, as have such invertebrates as insects or protozoans.

In a 577-page monograph on the amphibians, Noble (1931) devotes 46 lines to "Economic Value," mostly referring to the harvest of edible frogs' legs in the United States. Indeed, aside from their being a dietary element of vanishingly small importance to human beings in various parts of the world, amphibians could suddenly be eliminated entirely from the world and the material aspects of human welfare would be almost unaffected. Almost, for certain brightly colored tropical frogs of the genus *Dendrobates* produce cutaneous secretions that are sufficiently toxic to be used as arrow poisons by natives of Colombia; dried frogs and toads have been used medicinally in China and Japan (Henderson, 1864; Dunn, 1923); and, of course, as anyone who has studied introductory biology knows, there is a brisk trade in the United States in pickled and live frogs (usually *Rana pipiens*) to be used for educational purposes. Here the amphibian—or more properly, the frog—occupies a class by itself, having traditionally represented an entire group of vertebrates in American biological pedagogy.

Amphibians and reptiles are major vertebrate groups in which there are neither domestic species nor paradomestic species, such as are found among mammals, birds, and fishes. Paradomestic species are animal "weeds"—species that are closely dependent on environments uniquely created by human activity, without yet having been taken in deliberately or absorbed as useful entities into the human household. Examples of paradomestic species would be the Norway rat (*Rattus norvegicus*), some of the domestic cockroaches (*Blatta orientalis, Periplaneta americana*), and the house sparrow (*Passer domesticus*), at least over part of its range. In domestication or paradomestication, some clear and significant relationship to man is always involved, though it may consist of nothing further than a strong positive aesthetic response, as is true with the goldfish and the canary. These clear and significant relationships may be less common among amphibians and reptiles than among other vertebrates.

The foregoing is not promising ground on which to hunt for pests or potential pests; in fact, I have been unable to locate a good example of either. To illustrate how difficult it is to cite examples, I might mention the following: There is a large toad, *Bufo marinus*, native to South America, a predator upon certain scarabaeid beetles that are serious pests of sugar cane. It has been carried to Puerto Rico, Haiti, Guam, Hawaii, the Solomon Islands, Australia and other places where sugar is grown, and where it functions successfully as a biological control of the beetles that damage the cane (Cochran, 1961). According to the Australian News and Information Bureau (no publication date; received by me in 1967): "The giant toad, which can grow to eight inches long and five inches wide, was introduced to control a beetle that damages sugar cane. It is now a bad pest, killing useful insect life and many birds." Further, ". . . the giant American toad was introduced into Queensland in the 1930's to combat the sugar-cane beetles. But it, in turn, became a pest." Unfortunately, I have been unable to find further references to this situation, which is of some interest here because of the exotic status of this toad in Australia. The example demonstrates that amphibians are no different from other organisms in their potential for behaving in unexpectedly unpleasant ways when introduced into communities where they are not native.

Men have so seldom shown interest in controlling amphibian populations that there is nothing to draw on by way of experience with various methods of control. The susceptibility of amphibians to insecticides will be discussed briefly later. Inadvertent control through other deterioration of habitat has probably begun to effect great changes in amphibian faunas and amphibian population sizes over much of the more humid continental land surface of the earth. There is little documentation of the retreat of amphibians from settled, cultivated, developed, and finally urbanized land. Certain changes result as the population of human beings increases. Pollution, accelerated erosion and siltation, drainage of land for cropping, lowering of water tables, and land-filling for building—all of these affect the quality or availability of surface water. Amphibians, owing to their dependence on free water, probably respond to those changes more promptly and more strikingly than other small terrestrial vertebrates, perhaps to about the same extent as fishes. Concurrent changes—simplification of plant communities, elimination of anything but cultivated vegetation, reduction in invertebrate faunal diversity, and other similar alterations—attack amphibians on dry land. Because they have no economic importance, the amphibians are also without historian: if vast herds of salamanders had, like vast herds of bison, been driven from the plains by arriving settlement and agriculture, we would not be certain to know of it. We know little now about responses of amphibians to particular patterns of change in land use, excepting a very few cases (like that of *Rana fisheri* in Nevada) of outright extinction (Matthiessen, 1959).

REPTILES

Crocodiles

It is also difficult to find species among the reptiles that can be properly described as pests. However, certain of the larger crocodilians may injure or kill persons or prey upon domestic animals such as dogs, swine, goats, and poultry. Also, a considerable number of species of venomous snakes are capable of killing or seriously injuring persons. An element of drama in both relationships has seemingly tempted folklore to exaggerate their importance, but they are worth considering briefly.

There are 25 living species of crocodilians, widely spread in the tropics throughout the world. Of these, only two apparently cause people any serious difficulty. Both are Old World species of the genus *Crocodylus*. The Nile or African crocodile, *C. niloticus*, is found throughout interior tropical and southern Africa wherever there is suitable habitat, on Madagascar and some

of the offshore islands, and apparently was displaced in historic times from
the lower Nile and portions of the southern Mediterranean coast. The salt-
water crocodile, *C. porosus*, is a marine and coastal species ranging from Ceylon and the east coast of India throughout the Malay Archipelago eastward to
the Philippines, Fiji, and the Solomon Islands, and southward to northern
Australia, nowhere penetrating very far inland. A third species, the mugger,
C. palustris, of the interior fresh waters of India and Ceylon, has a sinister
literary reputation, probably undeserved, that follows from its habit of eating
corpses washed from burning ghats along the Indian rivers. In Ceylon it may
be dangerous to man (Deraniyagala, 1939).

Is there a green branch and an iron ring hanging over a doorway? The Old Mugger knows
that a boy has been born in that house, and must some day come down to the Ghaut to
play. Is a maiden married? The Old Mugger knows, for he sees the men carry gifts back
and forth; and she, too, comes down to the Ghaut to bathe before her wedding, and—he
is there.

KIPLING, *The Undertakers*

Crocodile attacks on persons always appear to be attempts to capture food.
The victim is seized and pulled into the water, where the crocodile attempts
to disable or dismember him by holding tightly with the jaws while throwing
its heavy body into vigorous twisting or thrashing motions. This technique will
detach limbs or twist or tear the victim into pieces. If the victim does not
escape, he is drowned eventually and the dismemberment continues until he
is reduced to manageable portions and swallowed. Attacks are apparently always made from the water, on people swimming, wading, trailing their hands
in the water, fallen from boats, or working at the edge, filling water jars, washing clothes, bathing, and so on. Stories of crocodiles foraging for victims on
land or in villages are not convincing; they are probably related to the habits
of certain species, particularly the mugger, of wandering overland in search of
suitable habitat during droughts. Injuries sustained by victims who have escaped
from crocodiles after being attacked are often massive, including serious lacerations, loss or dislocation of limbs or portions of them, and broken bones, particularly those of the limbs.

Whether or not a particular species of crocodile is a threat to men is apparently not a matter of size but of temperament. Some of the largest species of
crocodilians (*Crocodylus acutus, C. intermedius*) have no records as maneaters
(see Schmidt and Inger, 1957, and Pope, 1955, for information on sizes of
various species). The American alligator, one of the largest species of crocodil-
ian, cannot be considered dangerous to man (McIlhenny, 1935). Of the two
demonstrably dangerous species, *C. porosus* is larger (average length 12–14 ft)
than *C. niloticus* (average length 12 ft). Both are to be distinguished by their
aggressiveness or willingness to stalk and attack men. This aggressiveness is

perhaps related to the ordinary choice of food of the species. All crocodilians
are exclusively carnivorous, but their diets as adults range over considerable
variety, from small invertebrates, such as shrimps and crabs, through fish,
turtles, snakes, aquatic birds, to larger mammals, including those the size of
cattle (Schmidt and Inger, 1957). The largest prey must be dealt with by the
twisting-dismembering technique previously described. It is possible that in
certain species the most highly-evolved form of this behavior pattern is associated with a tendency to be attracted to larger prey, including man. Detailed
comparative studies of food habits and food-collecting behavior are missing
from the literature. Schmidt and Inger (1957) report that in Africa, the dispo-
sition of *C. niloticus* is variable from place to place. It is very dangerous in
some localities, where permanent stockades are built at the water's edge to
protect people filling water jars; elsewhere, where the same species is present,
precautions are not taken and few injuries sustained. No explanation is given
for these local differences, and apparently no attempts have been made to relate them to local differences in choice of food or in food-collecting behavior.

The significance of crocodile injury to human welfare in general must be
very close to nil. Useful statistics on mortality rates from crocodile injuries are
not available, partly because in places where rates are highest, records are most
fragmentary. Estimates that are encountered here and there, such as the sug-
gestion that 1,000 lives are lost annually on the lower Zambesi from crocodile
attacks (Earl, 1954), are probably exaggerations. Unfortunately, in the official
classification of causes of death* used by the World Health Organization (WHO)
and by most countries that compile mortality statistics in which deaths are
systematically reported by classified causes, deaths from crocodile attacks are
reported in Category E-928. This category also includes deaths resulting from
bites of any nonvenomous animal, including domestic ones, and deaths that
follow from being fallen on, stepped on, kicked, gored, or scratched by any
such animal, etc. Within this category, crocodiles are probably responsible for
only a small fraction of the deaths recorded, even in localities where crocodile
attacks are most frequent. The same comment must be made about losses sustained from destruction of domestic animals. There is undoubtedly a small,
steady, economic drain in particular areas where domestic animals share a
habitat with crocodiles, but insufficient data are available to describe it quantitatively.

Probably more species of crocodilians are predators on domestic animals
than on human beings, thus injury to domestic animals probably occurs over
a wider geographic area than injury to persons.

International List of Causes of Death. Published originally in supplement 1 of the Bulle-
tin of the World Health Organization, Geneva, 1948. Revised in successive editions of the
*Manual of the International Statistical Classification of Diseases, Injuries and Causes of
Death*, published by W H O.

Native populations exposed to risks of injury from crocodiles do not seem to have responded by making any systematic attempts to control crocodile populations. Indeed, the animal is often venerated as an ancestor. Sometimes there is ceremonial hunting. Frazer, in *The Golden Bough*, tells of the veneration and ceremonial execution of crocodiles on Madagascar, where the crocodiles are also sought for food, in the form of eggs or as juveniles or adults. The natives of the Irrawaddy delta esteem crocodile flesh and capture *C. porosus* with baited hooks.

The advent of Europeans with firearms resulted in greatly increased mortality among the crocodiles. They were destroyed for sanitary reasons, for sport, and also for their hides, which in certain species yield a leather of very high quality. Despite their formidable qualities, crocodilians are vulnerable to human predation. They may be caught with baited iron hooks thrown into the water; they are easily killed with rifle bullets (although apparently difficult to approach during the day, they may be hunted at night with lights); in the Americas they are harpooned or hooked out of burrows where they hide or hibernate (McIlhenny, 1935). At one time they were captured in large numbers in the hatching stage to be sold as pets in the United States. When intense hide-hunting pressure was leveled against the American alligator, it suffered great reduction in numbers. The combination of hunting and the deterioration of its habitat resulting from drainage and land filling might well have pushed it to extinction had it not eventually been accorded protected status.

Populations of tropical crocodilians appear somewhat better able to withstand hunting pressure. But these too are probably being reduced by excessive exploitation, particularly in South America, from which most of the "alligator" leather now used in this country originates. It is unlikely that there will ever be a problem of controlling crocodilians as pests, but not at all improbable that sooner or later they will have to be protected from excessive reduction in numbers.

Snakes

The venomous snakes are grouped into four families. They are so well distributed throughout the world that much of mankind lives under a small risk of being bitten by at least one species. The families, with summaries of their distributions, are as follows:

Elapidae (cobras, coral snakes, mambas, etc.): Americas, from Paraguay north to southern boundary of United States; Australia, southeast Asia, India.

Hydrophidae (sea snakes): Tropical Pacific and Indian Oceans; marine, with a few landlocked species.

Viperidae (true vipers): Most of Europe, Asia, and Africa; not in the New World or Australia.

Crotalidae (pit vipers, including rattlesnakes): Americas; southeast Asia through temperate Asia, Ceylon, to extreme southeastern Europe; not in Africa or Australia.

In addition to these, the large family *Colubridae* contains a number of venomous species in the subfamily *Boigidae* (the so-called opisthoglyph or rear-fanged snakes). All are considered harmless to man, with one exception—the African boomslang (*Dispholidus typus*), a dangerously poisonous species.

Although some species have a very potent venom and are technically able to injure men, the danger from hydrophid snakes may be ignored. These are caught and handled alive in large numbers in specialized "fisheries" on the Indian, Malayan, and Philippine coasts, where there appears to be but little hazard to the fishermen. Beyond this, no species of sea snake has a consistent record of willingness to bite men. The venom is apparently used primarily in the capture of fish on which the snakes feed, and it is not used in defense. Each of the remaining three families contains species known to be dangerous to men in certain circumstances.

The geographical distribution of the risk to a man of injury or death from snakebite is exceedingly complex. It obviously does not depend alone on the ranges of the various poisonous species or on their local population densities, but on other factors as well. For example, a given poisonous species will produce more or less potent venom and because of its characteristic average size, will be able to deliver relatively great or small quantities in a successful bite. Because of its temperament, it will be more or less inclined to attempt to bite a man under each of the various circumstances in which it may encounter one. Because of its capability, size, or habit, a typical attempt to bite may succeed or may fail to deliver an injurious dose of venom. All the above characteristics of the species of snake will vary with time of day, season, immediate environment, and, in individuals, with age, size, and perhaps with sex.

The person who is bitten contributes to the interaction. Whether an attempt is made to bite him will depend upon certain cultural characteristics he has: whether he tends to move about in certain kinds of vegetation at certain times of day or in certain seasons; on what kind of house he builds; and on what kinds of domestic animals he shelters. Whether the attempt is successful will depend on other variables: whether he wears shoes, or loose or tight clothing, or any clothing; or on other characteristics he has as an individual: whether he is old or young, agile or clumsy, deaf, blind, and so on. Whether or not he succumbs to a successful bite will depend on other extensive sets of cultural and individual variables: medical attention may be nearby or inaccessible; folk remedies may or may not be effective; his sex, age, and physical condition will vary, etc. Along with all the above variables specified, the geographical distribution of human population density may become a factor.

To provide a geographically characteristic index of risk of injury from snakebite, taking the above variables into account, would be a formidable task. This statement ought to be obvious. But apparently it does not always seem so, as when it is widely assumed that the mere presence of a venomous species of snake in a neighborhood makes it virtually certain that numbers of persons will be bitten and killed. In the United States, for example, it is entirely possible that larger numbers of people succumb to the exertions of actively hunting down and killing snakes as vermin than do to bites of the snakes themselves.

Meanwhile, the best geographic index of risk would be an actual crude or specific mortality rate. For reasons similar to those given previously in considering mortality from crocodile attacks, such rates are not easy to obtain. Deaths resulting from the bites of venomous snakes are not isolated in reports of WHO, where they are included in Category E-927 (consult the *International List*) along with those resulting from the bites or stings of all other venomous animals, including scorpions, spiders, centipedes, bees, wasps, jellyfish, etc. The same category is used in reporting official mortality statistics in the United States and in most other countries where deaths are systematically reported by classified causes. Within the category, snakes do not clearly contribute overwhelmingly to the total number of deaths reported; in Klauber's (1956) opinion, scorpions may kill more people in the United States than snakes do.

Another difficulty is that not all deaths or injuries listed as due to "snakebite" follow from the immediate toxic effects of snake venom. Often the snake escapes before it can be identified, or it is misidentified. Many nonvenomous snakes bite when molested, and many attacks by "poisonous" snakes are not that at all. In southern Wisconsin, for example, nonvenomous but aggressive water snakes of the genus *Natrix* are widely called "water moccasins" and are almost as widely thought to be venomous. The true water moccasin, a venomous crotalid, does not occur that far north. An occasional death from extreme fright or shock might be caused in particularly susceptible individuals from the bite or even sight of a nonvenomous snake. Some deaths attributed to the effects of snake venom may follow from injuries sustained by the victim during inept treatment of the bite (see Klauber, 1956, for an extensive discussion of this and related matters).

Generally speaking, deaths approximately assignable to bites of venomous snakes are relatively infrequent almost everywhere in the world. Table 1 is based on data of Swaroop and Grab (1954). To their estimates of mortality rates from snakebite for various countries, which apply to world conditions in about 1950, I have appended estimated total crude mortality rates for about the same period, for comparison. Swaroop and Grab (1954) and Allen (1949)

TABLE 1   Mortality Rates from Snakebite and from All Causes

	Mortality from Snakebite: Deaths per 100,000 of Population per Year[a]	Total Mortality: Deaths per 100,000 of Population per Year[b]
Australia	0.07	910.0
Burma	15.40	810.0[c]
Canada	0.02[d]	860.0
British Guiana	0.80[d]	990.0[e]
Ceylon	4.20	1090.0
Colombia	1.56[d]	1160.0[f]
Costa Rica	1.93[d]	1170.0
Egypt	0.20	1480.0[g]
England and Wales	0.02[d]	1140.0
France	0.06[d]	1300.0
India	5.40	1500.0
Italy	0.04[d]	990.0
Japan	0.13	890.0
Mexico	0.94	1560.0
Spain	0.02[d]	970.0
Thailand	1.30[d]	810.0[c]
Venezuela	3.10	990.0

[a]Data from Swaroop and Grab (1954).
[b]Data from *Encyclopedia Brittanica Yearbooks*, 1956 and 1967. Values given are for 1953 unless otherwise noted.
[c]Malaya, 1964.
[d]Includes bites or stings of all venomous animals.
[e]Venezuela, 1953.
[f]Peru, 1953.
[g]UAR, 1965.

agree in placing the worldwide mortality from venomous snakes at about 30,000 per year, or about one per 100,000 of population per year. Other data on mortality rates in various countries are reviewed by Klauber (1956). Taking great pains to resolve problems arising from poor data, he comes to the conclusion that in the United States (as of about 1956), about 1,000 persons per year are bitten by rattlesnakes, perhaps 1,500 are bitten by all venomous snakes, about 30 deaths result from rattlesnake bites, and somewhere between 40 and 45 deaths result from bites of all snakes, giving mortality rates of about 0.02 per 100,000 of population per year from rattlesnake bites and 0.027 from bites of all venomous snakes. A slightly more recent estimate of Parrish (1959) does not basically disagree with Klauber's estimate. Parrish examined the certificates for all deaths reported in the United States in category E-927 for the period 1950–1954—215 in all for the five years. Of these, 71 were due to snakebites (14.2 per year or about 0.01 per 100,000 of population per year). Of the remaining deaths, 86 were due to the stings of hymenopterous insects, 39 to spider bites, 5 to scorpion stings, and the rest to various other bites and stings. At least here, snakes were not the most important agents of deaths reported in category E-927.

Other Reptiles

In one or two respects other than the above, reptiles are reported to be obnoxious. The following list contains some examples, all of minor importance.

Venomous Lizards   The lizard family Helodermatidae, endemic to a small area in southwestern United States and western Mexico, contains only two species, which are closely related: the Gila monster (*Heloderma suspectum*) and the Mexican beaded lizard (*H. horridum*). They are the only known venomous lizards. The poison apparatus is morphologically unsnakelike. The venom glands are in the lower jaw (instead of above the upper jaw, as they are in all snakes) and they deliver venom through multiple ducts to a groove inside the lower lip, whence it is introduced from the mouth into wounds by means of grooves on the edges of short teeth in both upper and lower jaws (snakes have grooved or hollow poison fangs only in the upper jaw). Apparently the poison apparatus is seldom, if ever, used in capturing prey. Both species are slow-moving, largely nocturnal animals that feed on the eggs of reptiles and ground-nesting birds and on nestling birds and mammals, all of which are probably located by smell. When annoyed, however, an individual will swing its head from side to side, snapping its jaws. The snapping is poorly directed, but if the lizard catches something in its mouth, it bites firmly and holds on tenaciously, giving the venom time to enter lacerations made by the teeth.

A typical helodermatid bite, although painful, is ordinarily not fatal to an adult human being in good health. The number of deaths reasonably attributable to the bite of the Gila monster in the United States is exceedingly small, perhaps no more than a dozen altogether in all available records (Bogert and Del Campo, 1956). Many of those reported have been children or adults already debilitated by alcohol, alcoholism, or other conditions. Klauber (1956) reports for Arizona (the only state in which the lizard has ever occurred in any considerable numbers) one death from the bite of a Gila monster during the 22-year period 1929–1951, omitting 1937.

Vigorous and unnecessary persecution has reduced the range and depleted the numbers of the Gila monster in the United States. Since 1952 it has been protected by law in Arizona from molestation and from commercial exploitation in the live condition.

Predatory Lizards and Snakes   Some carnivorous reptiles are large enough to be troublesome around domestic establishments. Included here are the larger constrictor snakes (Boidae and Pythonidae) and the monitor lizards (Varanidae).

In some tropical species, the constrictor snakes are large enough to capture and swallow chickens, ducks, rabbits, dogs, pigs, and other mammals or birds, ranging up to weights of about 60 pounds. In Malaya, some pythons appear to be robbers of chicken runs. It is not unlikely that those species whose preferences for habitat near human beings make them the worst predators on domestic animals also contribute substantially to domestic rodent control. Folklore notwithstanding, these snakes are not a serious danger to man himself, although injuries are occasionally recorded. Schmidt and Inger (1957) refer to the "authentic" case of a 14-year-old Malay boy who was swallowed by a reticulate python (*Python reticulatus*) in the Talaud Islands, the only record of its kind I have ever seen. Encountered in the wild, these snakes will ordinarily attempt to escape. On occasion they will strike defensively, with the mouth wide open. The teeth are numerous and long; if the snake is large enough, terrible lacerations can be produced by such a strike.

The larger monitor lizards, in the Komodo dragon (*Varanus komodoensis*) reaching a length of 10 ft and weight of 300 lb, are also predators at times on domestic animals, particularly on poultry. The Malayan monitor (*V. salvator*) is particularly notorious. Monitors are restricted in distribution to the Old World tropics. All are strictly carnivorous. There are perhaps a dozen species in Africa, southeast Asia, and the East Indies large enough to behave as agricultural pests. Here again, however, the habit of foraging for food near human habitations may well imply that a species has value in rodent control. Monitors are not known to attack man.

Certain other large lizards have reputations for being destructive to young poultry or eggs. The common iguana (*Iguana iguana*), which ranges from central Mexico southward through Central America into the West Indies and the South American tropics, and which may reach a length of 6 ft, as well as the teju (*Tupinambis nigropunctatus*), a large teiid of tropical South America, are examples.

No data are available from which estimates might be made of economic losses from reptile predation, but they are probably extremely small.

Aside from the examples given above, other reptile predation is mostly supposition and rumor. Often, any animal that looks dangerous or destructive to an imaginative eye is considered *ipso facto* to be so. The common North American snapping turtle (*Chelydra serpentina*), for example, has a reputation as a serious predator on gamefish and on young waterfowl, and also as a danger to swimmers. This probably originates from its appearance and from its defensive behavior when molested, for there is no evidence for any of it. The animal is exceptionally inoffensive and retiring while at large: careful studies that have been made of its diet strongly suggest that it feeds primarily on vegetation and carrion (Lagler, 1943).

SUSCEPTIBILITY TO CONTROL

As yet no attempt has been made to control a species of amphibian or reptile by the indiscriminate application of a pesticidal chemical to areas of land or water. However, amphibians and reptiles are sometimes injured by commonly used insecticides.

Unfortunately, there is little information available on the susceptibility of particular species to injury from various pesticides, and there have been few studies of the responses of populations to blanket applications. Rudd (1964) summarizes some of the experiences with DDT: "With this chemical it appears that mortality in amphibians is severe at an application rate of 1 pound per acre and becomes catastrophic at 2 pounds or more per acre. . . . Comparable responses in reptiles begin at about 2 pounds per acre, probably becoming serious at about twice this rate." Heptachlor, applied at rates of 1.5 to 2 lb per acre for control of fire ants in Alabama, apparently brought about marked reductions in populations of various species of frogs, snakes, and lizards, perhaps eliminating some species from the treated areas (DeWitt and George, 1960).

A few studies, however, have shown that, as expected, reptiles and amphibians are able to accumulate various pesticides present in their food, as are many other animals (Hunt and Bischoff, 1960; Pillmore, 1961; Rudd, 1964).

Delayed effects appearing after reconcentration of pesticide taken up in foods such as insects, earthworms, or fish have been reported only infrequently for amphibians or reptiles, probably because they have not been systematically looked for (Herald, 1949; Logier, 1949). Ultimately, the biological effects of particular widely-dispersed toxic chemicals will depend in part on the abilities of organisms to become resistant to them. Resistance to a particular toxic compound can evolve rapidly in a treated population, which has been demonstrated in insects and microorganisms. Vertebrates will no doubt respond in exactly the same way. Generation times of typical vertebrates are greater than those of insects and bacteria; thus it is to be expected that resistance may take longer to appear among vertebrates than it has among insects. Where compounds are chosen for their effectiveness on insects, the intensity of selection against susceptible vertebrates may be less than that against susceptible insects, further delaying the appearance of noticeably resistant vertebrate strains. Few species of vertebrates showing resistance have been discovered, but among them are cricket frogs (*Acris* spp.) in bodies of water adjacent to cotton lands heavily treated over a long period with DDT and, less often, with toxaphene, methyl parathion, and endrin, in the Mississippi Delta (Boyd *et al.*, 1963; Ferguson, 1963a, 1963b). Evidence for resistance lies in the ability of cricket frogs from populations chronically exposed to DDT to survive exposure to water contaminated with insecticides and lethal to individuals taken from populations of the same species previously unexposed to DDT. The same frogs had apparently acquired a crossresistance to aldrin, although previously unexposed to it.

Interesting as these findings are, they do not suggest that pesticide resistance will eventually protect amphibian or reptile faunas from depletion in areas chronically treated. When insects are the primary pests and vertebrates the adventitious victims, one might assume that the former will usually adapt to the pesticide faster than the latter. Significant resistance in the pest requires a change in the treatment, which renews the attack on the adventitious victim as well, but under conditions where more harm is likely to be done to it than under the previous treatment. For example, its population may not have recovered its previous size, at least not to the extent that the pest itself has.

SUMMARY

No species of amphibian is a pest, and the nature of the group makes it unlikely that circumstances will arise to cause a species to behave as a pest. There is, therefore, no need to consider seriously methods that might be used to control amphibians. However, their susceptibility to pesticidal chemicals and their great dependence on susceptible invertebrate (particularly insect) foods make it probable that they will suffer greatly as innocent bystanders from programs of chemical insect-pest control. Assuming that use of various insecticides will continue at present or increased intensity, and that no substantial changes are made in the techniques of application, it is reasonable to expect a depletion in the world's amphibian fauna in the near future.

Except for two well-defined interactions between particular reptiles and man (in both of which the harm done to man is quantitatively unimportant), the situation with reptiles is similar to that of amphibians. With reptiles, too, it is unlikely that circumstances will cause them to behave as serious pests. The danger of venomous snakes to man will tend to decrease as the habitat of human beings becomes more and more artificial, although perhaps with transitory fluctuations in certain parts of the world. In the United States, the rapid diffusion of urban populations into suburbs and the increasing frequency of ventures into the countryside by inexperienced persons, may temporarily raise mortality rates from snakebites. Generally speaking, however, the snakes will disappear from the suburbs as rapidly as people arrive, owing to devastation of their habitats and disappearance of their food supplies. No highly specific and effective means has been discussed for eliminating dangerous snakes selectively from areas where they are likely to interfere with outdoor recreation.

Although more robust in the presence of insecticides than amphibians, reptiles are also susceptible and are likely to respond generally in the same way, with a decline in species diversity associated with increasing diffusion and accumulation of insecticidal chemicals in the environment.

As for crocodiles, it is not likely that their present status as pests will change for the worse in Africa or in Asia or that they will become pests in other parts of the world. The ready reduction in population density of crocodilians that can be achieved by hunting, where there is adequate incentive to maintain the pressure, suggests that regulation of crocodile numbers will not be difficult where it is necessary. Perhaps the high price of crocodile leather will cause them eventually to die out.

REFERENCES

Allen, F. M. 1949. Venomous bites and stings. J. Amer. Med. Ass. 139(9):616.

Bogert, C. M., and R. M. Del Campo. 1956. The Gila monster and its allies—the relationships, habits and behavior of the lizards of the family Helodermatidae. Bull. Amer. Mus. Natur. Hist. 109(1):1–238.

Boyd, C. E., S. B. Vinson, and D. E. Ferguson. 1963. Possible DDT resistance in two species of frogs. Copeia 1963(2):426–429.

Cochran, D. M. 1961. Living amphibians of the world. Doubleday, Garden City, New York. 199 pp.

Darlington, P. J. 1957. Zoogeography. John Wiley & Sons, New York. 675 pp.

Deraniyagala, P. E. P. 1939. The tetrapod reptiles of Ceylon. Ceylon J. Sci. 1:1–412.

DeWitt, J. B., and J. L. George. 1960. Pesticide-wildlife review, 1959. Bur. Sports Fish. Wildl., U.S. Fish & Wildl. Serv. Circ. 84 rev. 36 pp.

Dunn, E. R. 1923. The salamanders of the family Hynobiidae. Proc. Amer. Acad. Arts & Sci. 58:445–523.

Earl, L. 1954. Crocodile fever. Collins, London. 255 pp.

Ferguson, D. E. 1963a. Mississippi Delta wildlife developing resistance to pesticides. Agric. Chem., Sept. 1963. 3 pp.

Ferguson, D. E. 1963b. Notes concerning the effects of heptachlor on certain poikilotherms. Copeia 1963(2):441–443.

Henderson, J. 1864. The medicine and medical practice of the Chinese. J. Roy. Asiat. Soc. N. China Br. 1 (n.s.):21–69.

Herald, E. S. 1949. Effects of DDT–oil solutions upon amphibians and reptiles. Herpetologica 3:117–120.

Hunt, E. G., and A. I. Bischoff. 1960. Inimical effects on wildlife of periodic DDT applications to Clear Lake. California Fish & Game 46(1):91–106.

Klauber, L. M. 1956. Rattlesnakes. Univ. California Press, Berkeley. 1476 pp.

Lagler, K. F. 1943. Food habits and economic relations of the turtles of Michigan with special reference to fish management. Amer. Midland Natur. 29(2):257–312.

Logier, E. B. S. 1949. Effect of DDT on amphibians and reptiles. *In* Forest spraying and some effects of DDT. Canada Department of Lands & Forests, Biol. Bull. No. 2, pp. 49–56.

McIlhenny, E. A. 1935. The alligator's life history. Christopher, Boston. 117 pp.

Matthiessen, P. 1959. Wildlife in America. Viking Press, New York. 304 pp.

Noble, G. K. 1931. The biology of the amphibia. McGraw-Hill, New York. 577 pp.

Parrish, H. M. 1959. Deaths from bites and stings of venomous animals and insects in the United States. Arch. Intern. Med. (Amer. Med. Ass.) 104:198–207.

Pillmore, R. E. 1961. Pesticide investigations of the 1960 mortality of fish-eating birds on Klamath Basin wildlife refuges. U.S. Fish & Wildlife Serv., Wildl. Res. Lab. 12 pp.

Pope, C. H. 1955. The reptile world. Alfred Knopf, New York. 325 pp.

Rudd, R. L. 1964. Pesticides and the living landscape. Univ. Wisconsin Press, Madison, 320 pp.

Schmidt, K. P., and R. F. Inger. 1957. Living reptiles of the world. Hanover House, Garden City, New York. 287 pp.

Swaroop, S., and B. Grab. 1954. Snakebite mortality in the world. Bull. WHO, United Nations 10(1):35–76.

CHAPTER 4

Birds in Pest Situations

Singly, in small groups, or in large aggregations, birds struggling for survival conflict with man. Much of the adverse interaction is prompted by the activities of man, for he inadvertently encroaches on natural bird habitats and creates "attractive nuisance" habitats. It is irrational to think that the needs of birds take preference over those of man, but cannot both be achieved compatibly?

An accurate assessment of losses due to birds is difficult to obtain, but we can examine the estimates and data that are available. Losses occur largely in the destruction or contamination of food and fiber by birds. Economic losses in agriculture would seem to be easy to compile, but source material is annoyingly vague. Data on crop loss are difficult to assess because crop damage is usually concentrated in limited areas or even on single holdings. Widespread losses are less common, well known, and publicized. One estimate shows that crop loss attributable to birds is between $50 and $100 million annually (Miller, 1967). Some general figures on losses for several states are given in Table 1.

Costs of prophylactic activities to safeguard human health are virtually impossible to obtain. Aesthetic degradation and discomfort caused by birds are subjective evaluations that (except for commercialized recreation) defy reduction into monetary terms. Public and private indignation aroused, for example, by starling roosts in residential areas or on hotel buildings is often disproportional to the actual costs of prevention or control.

Since the crash of a jet airliner at the Logan Airport in Boston, in which 62 human lives were lost, the effect of birds on aircraft has been scrutinized by commercial airlines and by the military. This tragedy was caused by a flock of starlings that was drawn into the jet engines and caused them to fail. The loss of such a plane could cost more than a million dollars. Jet engines fouled by birds require expensive repairs or replacement.

Less serious, but still inconvenient and costly, are the protection and sanitation of public buildings, monuments, and statuary fouled by pigeons, English sparrows, and starlings. To the urban home owner, a neighbor's pigeons can be annoying and unsanitary.

The objective of bird control is to alter the pest situation by adjusting bird, man, or habitat to alleviate the adverse interaction between man and bird. Such adjustment must be made with the minimum amount of destructive activity affecting each component of the environment.

In order to achieve the best ecological solution through an action program, the pest situation must be analyzed. We must first ascertain the *kind and amount of damage*. For example, the corn in the field is partially eaten, resulting in a loss of harvestable corn. This depredation can be reduced to bushels per acre and the *monetary loss* calculated in dollars per acre. The same can be done for rice, fruits, vegetables, seed, etc. In the case of grain consumption in feedlots, the calculations are more involved. One realistic appraisal (Besser *et al.*, 1968) uses number of days spent at feedlots in winter X percent of diet obtained from cattle feed troughs X daily consumption capability (in lb) X size of bird flock. It was concluded, using 1966 feed costs, that the annual loss to cattlemen using feedlots was $84 per thousand starlings and $2 per thousand redwings. The economic factor is no small part of the motivation that governs control intensity and methodology.

An appraisal of loss in the Corn Belt was detailed by DeGrazio (1964): "The average corn loss [due to blackbirds] in the 94-square-mile study area for 1961 was 4.81 ± 3.03 bushels per acre; the 1962 loss, 2.38 ± 1.37 bushels; and the 1963 loss, 2.48 ± 1.31 bushels. Using a figure of $1.00 per bushel of corn, the monetary loss was about $41,000 in 1961; in 1962, about $20,000; and in 1963, about $25,000." The percent of the overall crop loss was not given, but on one test area 43 percent of the unprotected corn was lost.

Geographic location is a factor in appraising control procedures. For example, grape crops in both New York and California are likely to be attacked by fruit-eating birds. Techniques of protection for fruits are often interchangeable between locations, as are devices and schemes for protecting cereal grain.

The *season of conflict* may be related to control or prevention of pest situations. Sandhill cranes damage grain crops in the fall, pheasants pull corn in the spring, woodpeckers damage pine cones and seeds in the late summer, and mergansers depredate trout streams and hatcheries primarily in winter.

The *duration* of the impact is also a factor in control programs. Blackbirds

TABLE 1 Some Bird Depredation Losses

Species	Loss	Amount	Location	Season	Reference
Pintail	Grain (40% of 10,000 acres)	$216,000	California	FWS	Day, 1944
Mallard	Corn	$4,500/day	Colorado	FW	Wagar, 1946
American widgeon	Lettuce	$60,000 in 1 week	California	SFW	Horn, 1949
Blackbirds	Rice	$1.5–3 million/yr	Arkansas	SpF	Williams, 1960
Water birds	Fish fry	ca. $228,000/yr	38 states	SpSFW	Lagler, 1939
Blackbirds	Corn (Sandusky county)	$866,000 in 1 yr	Ohio	SF	Kottman, 1967

and starlings in feedlots create a winter-long problem in some areas, whereas
songbird attacks on fruits may only cover a span of two to three weeks, and a
transient hawk may create a pest situation lasting only two or three days.

From the analysis of these five factors—the kind and amount of damage,
the financial loss, the geographic location, the season of conflict, and the duration—will come the decision as to whether control is imperative, helpful, or
not required.

AREAS OF CONFLICT

Areas of conflict exist when birds

1. Consume or destroy foodstuffs,
2. Cause economic losses (nonfoodstuffs),
3. Jeopardize safety in aircraft,
4. Become carriers of diseases affecting man or the organisms over which
   he exercises husbandry, and
5. Negatively affect man's comfort, aesthetics, or sporting values.

Each area of conflict has its special aspects for identifying the problem and
its control methodology. It is impossible to list all levels of pest interaction,
but selected examples should serve to provide reference points for planning a
program of prevention or elimination of the pest situation.

BIRDS AND FOODSTUFFS

Today, more than ever before, the need for food is acute.

One of the major pest situations occurs in the cereal grain growing areas of
the Canadian prairie provinces and in the northern prairie states of the United
States. Here, waterfowl breeding marshes are often adjacent to grainfields.
The grain harvest coincides with the period when the current year's crop of
young ducks join with the adults prior to the southward migration. Waterfowl
aggregations "stubble feed" in grainfields, and in some areas as many as 40,000
stubbling ducks were found in a 100-square-mile grain-growing area (Hammond, 1961).

The harvesting procedure used in this region is to cut the grain in swaths
and to leave it on the ground to complete the ripening process, after which it
is picked up by a combine and threshed in the field. This method permits even
ripening and avoids kernel loss through shattering or seed drop. But grain thus
swathed is vulnerable to stubbling ducks. If rain causes flooding, the grain will

remain in the field for an extended period and the entire crop may be eaten or trampled into complete loss. Although such depredations do not occur in all grain-growing areas, large areas near marshes suffer catastrophic losses.

So serious have these losses become in recent years that two major relief programs have been undertaken. The first and obvious one is to chase or entice the birds away from the downed grain. Fright devices include gunfire, firecrackers, scarecrows, acetylene exploders, etc. (Often this drives birds to a neighbor's fields.) Or, the harvested grain is spread in fields near the marshes to entice the birds from the swathed fields. Both practices are costly and not completely effective. A second possibility is to provide insurance against crop damage. The provincial government of Saskatchewan has underwritten the crop-loss insurance program in part with monies from sportsmen's license fees (Paynter, 1966). In the past 13 years it has accepted liabilities of $9,557,828 and paid losses of $745,255 (Table 2). Alberta has used a similar plan since 1961. The preventive and redemptive programs are not mutually exclusive; in fact, they reduce economic loss best when used in combination.

Rice is another major crop damaged by waterfowl, as well as by such other species as blackbirds and cowbirds. The two major rice-growing areas are in the lower Mississippi Valley and in California. The intensity of depredation is great in both rice-growing areas, but the time when such damage occurs varies. Horn and Glasgow (1964) report: "Little or no damage is done to rice in the South and Southeast in the fall. This is the reverse of the situation in California, where spring damage is negligible and fall damage has been severe." Thus all aspects of appraisal of the pest situation and the control methods will differ even though bird and food resource are the same.

Some forage and truck crops are also eaten by waterfowl. A primary conflict occurs when widgeons invade new seedings of these crops. Tender foliage is preferred by both coots and geese. Although leaf pruning of cereal grains by geese has been shown to increase yields (Quinn, 1952), puddling by ducks and geese in fields that are wet or flooded causes the soil to become very hard when the water is removed or is evaporated. Cultivation becomes difficult and plants are lost.

Soft fruits and vegetables are particularly vulnerable to birds, especially the passerine or perching birds. Giltz (1960), in a survey directed to state experiment station directors, recorded that tomatoes, melons, grapes, blueberries, strawberries, lettuce, and cherries were some of the main soft fruits and vegetables attacked by birds. In most cases control is difficult. Although protective screens were most successful for blueberries, for example, the cost and awkwardness of operation for commercial production discouraged the use of screening devices (Hayne and Cardinell, 1949). Various scare devices used in fruit crops have limited efficiency, since birds frequently become refractory

TABLE 2 Saskatchewan Wildlife Crop-Damage Insurance

Year	Total Liability	Premiums Paid	Number of Farmers Insured	Losses Paid
1953	$ 21,500.00	$ 1,065.25	20	$ 2,387.75
1954	34,310.00	1,715.50	30	8,566.50
1955	188,761.00	5,677.53	144	14,725.37
1956	460,295.00	9,456.12	245	40,017.50
1957	582,050.00	11,626.12	335	59,348.00
1958	717,093.00	14,348.60	432	69,650.87
1959	694,229.00	13,901.00	407	152,207.00
1960	1,665,600.00	30,312.00	822	60,715.00
1961	495,557.00	9,911.00	285	24,417.00
1962	641,350.00	12,837.00	292	32,045.00
1963	1,025,000.00	20,574.00	422	44,786.00
1964	742,465.00	14,849.00	319	92,500.00
1965	2,289,618.00	36,909.00	642	143,890.00
	$9,557,828.00	$183,182.12	4,395	$745,255.99

Source: Paynter, 1966.

to such devices (Seubert, 1967). Nut trees and fig orchards are also subject to bird-pest situations.

Starlings and blackbirds in western feedlots do not affect the cattle directly, but consume food to a point where serious financial losses accrue. The pest situation lasts throughout the year, but is most serious during the nonbreeding period when flocks congregate in feedlots.

With regard to pest situations involving avian predators, careful field study supports the position that raptors (birds of prey) are rarely detrimental to human welfare. One of England's most knowledgeable naturalists, Col. R. Meinertzhagen (1959), states in his book *Pirates and Predators*: "The sportsman, the egg-collector, the bird-lover, bird-photographer and bird-watcher are all greater disturbers and destroyers of wild life than any wild predator, who after all is going about his lawful business—survival."

One of the best studies conducted on raptors and their prey (Craighead and Craighead, 1956) shows nongame food varied from 50 percent of the spring and summer diet to 95 percent in fall and winter. The domestic chicken comprised less than one percent of their total diet from February to August on typical midwest farmlands. This does not imply that hawks and owls do not take poultry; they certainly do, but it is unusual rather than commonplace. If a raptor becomes habituated to the poultry yard, it should be eliminated. However, this should not be the occasion to set out a dozen pole traps or to shoot all hawks on sight. Protection laws now make this clear.

Fish culture in private and public hatcheries is under constant harrassment and occasionally suffers serious loss from such birds as herons, egrets, king-fishers, and mergansers (Lagler, 1939). The control efforts are often direct and stringent, and more often not a matter of public record.

BIRDS AND OTHER ECONOMIC LOSS

Certain forest birds normally feed on seeds, buds, twigs, and leaves or needles of trees, usually with no harmful effect. Only when concentrations of either birds or food trees occur do such feeding habits result in obvious damage. The pine and evening grosbeaks often congregate in winter and move south into re-gions of reforestation through softwood plantations or monotype seeding of conifers. Here damage to terminal and lateral buds can be severe. Grosbeaks eat buds of coniferous and some hardwood trees, but neither the pine or eve-ning grosbeak is an obligate or even strongly preferential feeder. Both grosbeaks otherwise feed on a wide variety of tree species.

"Birds are more of an influence in seed consumption than is generally be-lieved" (Smith and Aldous, 1947). Several studies on tree seeds eaten by birds list sparrows, juncos, blackbirds, towhees, mourning doves, and grosbeaks as the common offenders. In all, 37 species are known to eat conifer seeds. Forest seeding is expensive, and the results are seldom uniform in any one time or place. Natural viability of seeds, inadequate moisture, seed destruction, or poor seed bed may contribute to lack of seedling production.

Sooty grouse have also been known to clip buds and needles from conifers, particularly Douglas fir on the West Coast (Lawrence *et al.*, 1961). Again the pest situation is confined to grouse concentrations (not an abundant species) in places where young trees have been planted. There is no special control technique for this kind of damage since the time and place of interaction is not predictable, nor is it readily detected in more remote areas.

A more recent kind of damage to older pine plantations in the Midwest is the destruction of leaders and lateral branches at the tops of the trees by roost-ing crows. As more of these plantations become mature, crows prefer them to deciduous groves because of the protective nature of the foliage. Thus winter roosts of crows on the northern edge of the crow range are vulnerable to me-chanical damage caused by the weight of the roosting birds. Discouraging a roost by scaring or shooting as it begins to build up in late summer may reduce the number of crows or cause the roost to move, but once established it is vir-tually impossible to alter roosting behavior, particularly if the roost is large.

In the Northwest, holly is grown and sold during the Christmas holiday period. The holly groves make ideal roosting areas for starlings, whose drop-pings ruin the leaves for marketing.

Throughout the country, woodpeckers damage wooden utility poles. Poles are hollowed for nesting sites, food-storage places, or perhaps just as "play" activity. Utility companies must replace damaged poles or suffer more serious loss if poles should break during a storm. Such added costs are passed on to the public. Perhaps the most enlightening appraisal of the woodpecker problem on utility poles is that of McAtee (1911). His damage-strength chart for wooden poles is the only one we were able to find. Dennis (1967) indicated that chemically-treated posts were more vulnerable than those untreated. Attempts to prevent damage to poles are graphically summed up by Benton and Dickinson (1966): "After 5 years of testing some of the smelliest and stickiest compounds known to man, however, no satisfactory chemical repellent had been found. The woodpeckers hacked happily through everything on the menu and even ruined the aluminum tags used to distinguish which chemical was being used on each pole."

Power companies are also plagued with short circuits caused by birds nesting on poles, flying into wires, perching on wires, or by connecting wires with streams of excrement. Power failures are a financial burden to industry and an inconvenience to the public.

In large warehouses, broken windows permit access to roosting and nesting birds (e.g., house sparrows, starlings, and pigeons). The fouling of stored goods with droppings or excess nesting material causes financial losses.

Fish-hatchery operators are in constant battle with fish-eating birds such as herons, kingfishers, mergansers, terns, gulls, and on occasion, ospreys and eagles. There is also a conflict between operators of game farms and raptorial birds. Control is tantamount to extremely effective baiting and destruction of the decoyed victims. This kind of control data is difficult to obtain and to appraise. The magnitude of the control effort or the benefits thus derived is unknown.

Golden eagles are predators on sheep and goats to such an extent that skillful husbandry is required to prevent losses of newborn animals. Claims have been made that 50 percent of a lamb crop and 90 percent of a kid crop could be lost to "Mexican" eagles (golden eagles) in Texas (U.S. House of Representatives hearings, 1966). A predator-control report in table form accompanied this testimony, where it was shown that of 5,292 actually observed depredations, only 12 (less than 1 percent) were attributed to eagles. Of these one was a sheep and 11 were goats.

There is no denying that losses do occur through eagle predation, but one gets the impression that such loss is exaggerated. This is attested to in part by the report of Dr. Walter Spofford (U.S. Senate hearing, 1962), who surveyed the golden eagle problem for the National Audubon Society. He found that ". . . there was very little real evidence of eagle depredations in lambs or kids. . . ."

In spite of federal legislation intended to protect the bald and golden eagles, the governor of any state can at any time request permission to kill eagles and "... the Secretary [Department of Interior] *will* authorize such taking [of golden eagles] without a permit in whatever part or parts of the State and for such periods as he determines necessary to protect those interests." [11.3(b) Federal Register, Part 1, vol. 28, no. 23, Feb. 1, 1963.] The demonstrated need in the case of golden eagles is left to stockmen and governors. The Bureau of Sport Fisheries and Wildlife issues the permit to kill eagles but its staff has, according to the law, no voice in the basic decision.

Perhaps the best appraisal of the eagle-sheep conflict is presented in the *Congressional Record* for the Senate (1962). This document referred to earlier records of the proceedings of a hearing on the "Protection for the Golden Eagle" before a subcommittee of the Senate Committee on Commerce, chaired by Senator Ralph W. Yarborough of Texas. Summarization of the testimony is difficult, but these points emerge:

1. Golden eagles do indeed kill lambs and kids.
2. Indiscriminate killing of eagles for sport or profit has not eliminated stock losses over the sheep ranges of Texas.
3. Only free-ranging sheep (as opposed to herding, particularly at lambing time) suffer serious eagle predation.
4. As with all predator problems, occasional control in time and place may be necessary, as is the need to adjust the animal-husbandry practice to prevent the conflict of a wild and a domestic animal.
5. Indiscriminate killing of eagles (or any animal) involved in a pest situation is unnecessary and unjustified.

One encouraging aspect of the difference of opinion on the role of the golden eagle as a predator is that two opposing points of view will be focused on the problem by mutual consent and in cooperation. The National Audubon Society and the National Wool Growers' Association aided by the Bureau of Sport Fisheries and Wildlife will work jointly to learn "how to protect sheep ranchers' livestock without endangering the survival of the golden eagle" (Boardman, 1967). Conservation has taken another step forward.

The rationale that exempted the hawks, owls, and eagles from the protective blanket of the Migratory Bird Treaty Act of 1936 is now outdated by research and by society's recognition of the role these birds play and have played in our culture. This awareness is attested to by many states that have enacted protective legislation for these birds in the absence of federal laws to do so. In recent years some of our raptorial birds have undergone catastrophic reduction in numbers—the peregrine falcon in particular (Hickey, 1969). In the legal hiatus now existing, dealers in wild animals accentuate the problem by offering for sale hawks, owls, and eagles, including the rare peregrine falcon.

It is proper and timely that all raptorial birds be protected soon by the Migratory Bird Treaty Act; but failing inter-American acceptance, a federal law that will provide adequate protection is needed.

BIRDS AND AIRPLANES

When airplanes collide with birds in flight, the damage can cause structural and engine failures, often resulting in fatal crashes. Aircraft surviving air collisions require expensive repair.

In 1960, 62 persons died in a bird-caused crash of a commercial jet plane at Boston's Logan Airport. Similar collisions have occurred with gulls, a sandhill crane, whistling swans, a great blue heron, a snow goose, and other unidentified birds (U.S. Air Force, 1966) in the United States. In Canada, waterfowl and gulls are of prime concern, but the flocking of small birds does not appear to be as serious a problem, as is indicated by the birds identified in Canadian air strikes (Solman, 1966).

The air-strike pattern for Air Canada illustrates that although the number of collisions has declined over the span of years, the monthly pattern remains the same, generally coinciding with the bird migration periods in spring and fall. Approximate cost figures to repair or replace Canadian aircraft damage as reported by the Royal Canadian Air Force and the Department of Transport show that costs vary annually for the commercial carrier Air Canada but have been declining in recent years—from a high of nearly $350,000 in 1961 to about $75,000 in 1966. Solman states: "We believe the hazard to aircraft landing at and taking off from Canadian airports has been substantially reduced through habitat modification and bird dispersal activities recommended by the committee" (National Research Council's Associate Committee on Bird Hazards to Aircraft).

The picture in the United States differs only in magnitude, although the available statistics are less precise, particularly in completeness and in details of reporting. Perhaps the most edifying data are those of Seubert (1965, 1966). The Air Transport Association reported 1,255 strikes for the period March 31, 1961, through June 30, 1965. During the past two years the percentage of strikes by jet aircraft has increased slightly while the turboprop aircraft has changed little, and that involving piston aircraft has decreased. The data summarizing the picture of air strikes for commercial carriers are presented in Table 3.

It is not easy to identify the species of bird hit in air strikes, but the Cana-

TABLE 3   Frequency of Reported Bird-Plane Strikes at Airports Handling 70,000 or More Carrier Operations, 1962–66

Airport	Air Carrier Operations	Number of Strikes	Average Number of Strikes per 10,000 Operations
Atlanta	946,874	39	0.41
Baltimore[a]	73,228	4	0.55
Boston	734,807	50	0.68
Chicago (O'Hare)	2,000,666	89	0.44
Cincinnati	386,641	10	0.26
Cleveland	581,585	11	0.19
Dallas	769,544	9	0.12
Denver	479,795	47	0.98
Detroit	803,676	86	1.07
Honolulu	444,813	8	0.18
Houston[b]	250,350	11	0.44
Kansas City, Mo.	404,831	44	1.09
Los Angeles	1,445,875	51	0.35
Memphis, Tenn.[c]	71,339	5	0.70
Miami	812,528	34	0.42
Minneapolis	412,474	20	0.48
New Orleans	391,111	15	0.38
New York (Kennedy)	1,658,051	109	0.66
New York (LaGuardia)	614,575	20	0.33
Newark	837,842	73	0.87
Philadelphia	635,279	41	0.65
Pittsburgh	557,822	18	0.32
San Francisco	955,958	67	0.70
San Juan[d]	145,964	1	0.07
Seattle	373,243	31	0.83
St. Louis	494,963	13	0.26
Tampa[e]	142,910	3	0.21
Washington (National)	1,061,114	62	0.58

[a]1 year (1966)   [c]1 year (1966)   [e]2 years (1963–1964)
[b]3 years (1964–1966)   [d]2 years (1965–1966)

Source: Seubert, 1968.

dian report (Solman, 1966) lists 29 and the U.S. report (Seubert, 1966) 59 species. Often all that is recorded is a size description of the bird hit or a general category like "sparrows" or "sea gulls." The cases involving identified species cover a wide spectrum of the bird group including individuals such as the horned grebe and pine siskin.

Military personnel and equipment are also endangered when aircraft strike birds. In 1965, 839 strikes were recorded by the Air Force; most were below 1,000 feet. Birds were drawn into the engine in 109 cases and 75 engine replacements were necessary, at a cost of four or five million dollars (U.S. Air Force, 1966).

One discouraging aspect in appraising the problem of bird-aircraft collisions is that the recording of such encounters by the military and commercial carriers has not been uniform. Pilots have failed to report strikes that were later recorded by maintenance personnel (Seubert; 1965, 1966). "The change [decreased reporting by pilots] is unfortunate, since reports filed by maintenance personnel seldom include certain important data (air speed, altitude, flight stage)." The Air Force report states that a 1965 change in reporting procedures increased the strike tally "nearly six times the number for the previous year."

Military aircraft in the Pacific are involved whenever airstrips are placed on islands that are traditional nesting grounds for such large sea birds as the Laysan and black-footed albatrosses. The danger has not been completely removed despite recommendations by several research groups.

The problem of air strikes is largely a function of airport location and construction. Runways built on or near ideal bird habitat bring birds and aircraft into conflict. Low, flat areas ideal for airports are frequently associated with water or marshland vegetation, which may be the breeding or roosting sites for large water birds or flocking, smaller, perching birds. The obvious approach to control is through habitat manipulation. Surface water can be drained from ponds and associated marshland to eliminate roosting places. Updrafts used by soaring birds can be minimized by altering vegetation or terrain. If crops are grown on marshland sites near airports, they must not be those that will attract birds. Often these open areas not occupied by runways have been planted to grass. Constant maintenance is required to keep the grass short enough to discourage small rodents, which could attract raptorial birds, as well as to prevent the accumulation of fuel for grass fires.

No manipulation is likely to be completely successful. The grass environment is ideal for earthworms, and after a rain the worms crawl out on the air strip. This food supply attracts gulls, thereby increasing the strike hazard; in addition, the crushed earthworms can cause aircraft to skid on the slippery runway (Solman, 1966). The elimination of municipal dumps near airfields could also limit some bird populations and reduce air strike potentials. Trees, brush, and nearby abandoned buildings are possible attractants for birds and should be removed. Reducing the environment to its least attractive condition to birds but compatible with airport maintenance may allow other dispersing activities such as noisemakers, distress calls, and sundry scare devices to reduce the interaction to a tolerable level. The maintenance of bird-free airports should be as standard a program as that for maintaining smooth landing strips or safe traffic patterns.

A recent and more sophisticated way to reduce air strikes is to avoid the times and places where encounters are likely to occur. The use of radar to determine where bird flights or concentrations are, relative to altitude and distance from airfields, will aid pilots in avoiding flight paths and movement patterns of potentially dangerous birds in flight. Closed-circuit television may also serve the same purpose on the airport site.

Little is said in the biological literature of engineering efforts to improve aircraft construction to minimize danger through air strikes. Solman (1966) pessimistically states, "To make a completely bird-proof aircraft would involve creation of a structure of such weight that flight would not be economically feasible." Could not a plane be built that would reduce crash hazard by 50 or 75 percent?

Occasionally nest debris and droppings from birds using aircraft repair hangars have fallen into dismantled engines. The inherent danger and financial loss resulting from such bird activity have necessitated control measures (Smith, 1966).

Long-term research into the fundamentals of bird behavior and ecology should be of great value. However, the problem of aircraft safety cannot wait, and, as an immediate approach toward its solution, experimentation with all remedial manipulations involving bird or environment should parallel fundamental investigations. Aircraft design involving increased engineering and construction costs must not be written off as impractical or untenable.

BIRDS AND DISEASE

Wild birds and their role in the transmission of disease to man and his domestic animals is an old problem. Pasteur studied chicken cholera in his Paris laboratory in 1880 and parrot fever in 1893. The infectious diseases that are transmitted from wild animals to man are now referred to as *zoonoses*. Herman (1966) says that "veterinary public health specialists list more than a hundred diseases of animals in the wild that can be transmitted to man."

Apart from the numerous ways in which bird-borne diseases affect the economic status of man's various animal industries, public health is also affected. Bird problems of zoonoses can be grouped into the following categories:

1. Birds producing aerosols and excretions of disease organisms.

An example of aerosol-carried disease is the human respiratory ailment first referred to as parrot fever (psittacosis) and now associated with other birds and called *ornithosis*. Workers in domestic turkey-processing plants were frequent victims of ornithosis, as were pigeon fanciers and squab producers. Before import limitations on cage birds and bird quarantines, pet shop owners and their customers became infected by sick parrots and other bird pets. Since this disease is airborne, a single visit to a pet shop could cause infection. Today such possibilities are relatively few, and if the disease is contracted, antibiotics will reduce its virulence so that deaths are uncommon. The number of cases of psittacosis (ornithosis) in the United States as reported by the National Communicable Disease Center shows a rise to nearly 600 cases in the mid-1950's, falling again to about one-tenth that number by 1966.

The aerosol may include dust-like particles of bird feces. Feces also carry *Salmonella*, the bacteria group that causes severe intestinal upset. Not only does this disease affect man, but it may cause wholesale die-off among commercial poultry. Feral pigeons, starlings, and house sparrows can transmit the disease organism from one poultry yard to another through contamination of their feet with poultry feces. Blood tests will determine which birds or flocks are infected, and sanitary measures including the destruction of infected birds will control the disease.

2. Birds acting as intermediate hosts for infectious disease.

The role of birds and man as intermediate hosts is important today because the key disease group is encephalitis (arthropod-borne). Mosquitoes and ticks transmit these diseases from one host to another. If one of the hosts is man, he can be crippled or killed. These viral diseases have a variety of types that infect both wild and domestic mammals and birds as well as man. The exact ecology of the virus and its animal environment is not completely understood, and it is presently being studied in various parts of the world. Through movements and migration, bird hosts may carry some of the arbovirus diseases for long distances. Control, however, is largely concerned with the arthropod vector and its habitat.

3. Birds becoming a reservoir for disease organisms.

A bird acts as a disease reservoir when it harbors organisms that will ultimately cause disease in man. Thus, as in the case of encephalitis, the bird probably keeps the supply of virus alive and available to the mosquito for short periods. The biology of a virus wintering-over in the physiological system of birds is still unresolved. Other short-period reservoirs may be the body feathers or feet (as in the transmission of *Salmonella* on the feet of pigeons) of the vector bird. Ascarid eggs and Coccidia oocysts on soiled feet or ectoparasites on plumage are additional potential sources of infection (Marsh, 1962).

The disease called histoplasmosis is caused by the pathogenic fungus *Histoplasma capsulatum*, which thrives in a medium of molding litter and bird feces. The fungus is not found in chickens or their fresh droppings (Price, 1966). It enters the body as an aerosol and lodges in the lungs, where a pneumonia-like condition develops. The spores of this fungus may also be ingested and lesions may develop in the alimentary canal, liver, or spleen (Price, 1966). Deaths have been recorded but human infection results in clinical histoplasmosis only after prolonged contamination or after massive doses are inadvertently taken. Preventing the establishment of bird roosts that would cause accumulated droppings, using a respirator, or wetting down a contaminated area will reduce the infection hazard.

Of the 100 diseases alleged to be transmittable from birds to man, only two or three are serious and widespread. This does not imply that none will ever reach epidemic proportions, and public health vigilance is constantly required. Required also is the understanding of disease ecology so that people are not frightened into attempting to eradicate birds to control disease. Only an ecological appraisal of the interaction of birds, disease, and man will lead to proper remedial or therapeutic action.

Domestic animals—poultry and hogs—are also victims of disease introduced by birds. The four species of known importance as disease carriers to domestic animals are house sparrows, starlings, pigeons, and herring gulls. These birds have caused epizootics in domestic animals mainly as contaminants of food and water (*Pasteurella* and *Salmonella* infections), as mechanical vectors (Ascarid eggs and Coccidia oocysts), and as biological vectors (leucocytozoan in ducks [Marsh, 1962]).

COMFORT AND AESTHETICS

Bird pest situations develop when birds affect man's comfort and aesthetic or sporting values.

Nuisance roosts of starlings are found within cities in trees or on buildings. The accumulation of droppings on sidewalks and homes beneath roost trees is unsightly, and policing is an expensive maintenance chore. Starlings have been known to kill valuable shade trees through intensive roosting, and the clamor of thousands of roosting starlings is offensive.

Fright techniques or mechanical devices have been used to reduce starling roosting problems in residential areas. In some cases the only permanent solution has been the removal of roosting trees (Bliese, 1959). An alternative to this drastic measure is a heavy trimming of the branches, which reduces the attractiveness of the trees to birds (Buckley and Cottam, 1966). However, this also reduces the value of the trees.

Starlings, pigeons, and house sparrows roosting on buildings have been controlled by contact toxicants, adhesive repellents, electronic shocking devices, sonics, excluding birds by nets or wire, by altering the slope of the roosting ledge, or by shooting.

House sparrows inhabiting ivy on the sides of buildings are a noise-nuisance

to man. One effective means of dispersing the roost is to apply a heavy stream
of water on successive nights.

Pigeons and English sparrows are guilty of fouling park statuary, outdoor
theaters, athletic stadiums, and other buildings. Even mallard ducks and gulls
can necessitate costly maintenance at boating marinas when they foul boats
at anchor.

If pigeons and house sparrows are attracted to a site by food, then strict
sanitation of the premises will discourage them. The public's inclination to
feed birds, especially pigeons, is difficult to control and can only be changed
by a slow educational process. This is also true for littering habits.

Nests of pigeons and house sparrows create a fire hazard when built in
wooden structures and maintenance problems when downspouts are obstructed.

Birds can affect man's sporting values by reducing the production of young
game birds and hence the harvestable fall crop. Two examples are the depredations of crows on waterfowl nests and magpies on nests of sage grouse (Martin,
1942).

CONTROL OF BIRD PEST SITUATIONS

As man continues to dominate the earth and develop a landscape that con-
forms to his economic demands, he creates monocultures which, in turn,
create potential pest situations involving birds. In such "new environments"
as feedlots, orchards, grainfields, and skyscraper jungles some species of birds
congregate in large numbers. A pest situation develops, and there is a cry for
control measures.

Once the need for control has been established and the feasibility of undertaking a control program ascertained, it remains to be determined how the
mechanics of the program will be put into operation to achieve the desired
results. To approach regulatory methods directly—trapping, shooting, poisoning, inhibiting reproduction—necessitates a knowledge of the animal's anatomy,
physiology, and behavior. The indirect approach—alter, shield, adapt, remove—
requires a knowledge of the animal's environment, including food, cover, and
competing and noncompeting organisms.

The ecological watchword in the mechanics of any control program should
be *specificity*. This implies that the control program affects only the target
species and that the program is directed only to those individuals within the
species that are in conflict with man's interests. It is perhaps easiest to attain
such specificity with control chemicals (e.g., lamprey larvicides), but the effects of such controls are rarely as limited as intended.

In the development of control activity, the first consideration must be with the natural controls of the pest animal. If we understand the ecology of natural regulations, control methodology will suggest itself, often in compatibility with both environment and animal. Materials, methods, devices are only adjuncts to the proper utilization of detailed knowledge of ecology and sanitation (Spear, 1966).

No one pest control technique can be considered a panacea for all situations (Besser, 1962). Many control techniques may be needed, with little assurance that they will be effective the following season. In four years of trying to control blackbird roosts, Bliese (1959) found that the efforts to control the birds in one year (with the obvious exception of tree removal) had no effect on blackbird behavior the next season.

Many variables attend a pest situation; for example, weather conditions influence bird damage to crops. Severe winter weather increases grain loss and contamination by birds at feedlots from November through March, especially during periods of snow cover. There also appears to be a difference in response to control measures among juvenile and adult birds. Siebe (1964) indicates that in summer juvenile starlings were repelled by frightening devices and biosonics, whereas in the fall and winter these same devices were not effective at feedlots where large numbers of adult birds were concentrated.

The practice of good sanitation is often overlooked in bird-pest situations. Some bird-pest problems could be controlled by the elimination of food sources, water supplies, or nesting sites.

The problem of bird-pest situations is not new; it dates back to man's first efforts to cultivate crops. At first, control methods largely relied on fright techniques; these were followed by lethal techniques. More recently, the emphasis has been on preventive techniques. Control efforts under each of these techniques have been of a mechanical, chemical, and/or biological nature.

FRIGHT TECHNIQUES

Fright techniques range from the scarecrow and gun to the automatic acetylene exploder and the plastic "whirler." Shotguns and .22 rifles have been used by farmers for years as direct control agents to frighten or exterminate. Shellcrackers in 12-gauge shotguns have been useful for controlling local bird-pest situations. Automatic acetylene exploders are effective for some control efforts but have a high initial cost and some operational problems. Exploders are more efficient in combination with the shotgun, and by varying the location and the interval of the explosion (Faulkner, 1962).

The general effectiveness of scare devices is directly proportional to the availability of alternative sources of food and to the proper application of the method and is inversely proportional to the size of the area to be protected.

In many cases they offer the only protection for grain crops, garden crops, fruits, and berries (Mitchell, 1960).

Chemical agents that frighten birds act through the nervous system. Soporifics put the birds to sleep or produce fluttering, flopping, and distress symptoms that cause other birds to leave the baited areas and stay away for long periods. The birds suffer little pain and are incapacitated for only short periods (Williams and Neff, 1966). Other fright-producing chemicals induce similar behavior but do not have an anesthetizing effect. The repellent must not be toxic or distasteful, and it should not discolor the marketable product (Zajanc, 1962).

The application of biological knowledge in developing fright techniques is best demonstrated in sonics. This technique has been employed successfully in frightening starlings and blackbirds from roosting sites by reproducing their distress calls. Europeans have used sonics, instead of drugs, with some success. Crows have been driven from their nests with distress calls on cold nights. The eggs chill, and a generation of crows is lost (Frings, 1964).

Birds have different dialects in different regions, and recorded calls are sometimes good for local use only. The effectiveness of alarm and distress calls probably differs among species. The distress call is effective with starlings because the birds are highly gregarious. The spacing of calls and the fidelity of sound reproduction are important to efficiency.

Electronic shocking devices have been employed in protecting statuary and ornamental architecture on buildings. There is also some evidence that birds in flight respond to radiations of electrical energy, but this is not thoroughly understood and needs further study (Williams and Neff, 1966).

Sticky repellents have been used to keep roosting birds off sheltered ledges on buildings, but sometimes these compounds deface the buildings more than the birds do.

LETHAL TECHNIQUES

Lethal techniques alleviate damage problems through a combination of killing and frightening. Man has always felt a sense of accomplishment in killing pests. The dead animals can be compiled in impressive statistics, but any long-term benefits are in doubt.

It is a biological axiom that animal increase can usually be depicted as a sigmoid curve, with the maximum rate of increase at half of the carrying capacity of the habitat. Therefore, the growth of the population is self-limiting as the population nears the carrying capacity of the habitat. This self-limitation occurs in part through the action of density-dependent factors. A partial reduction of a bird-pest species in any given year through either mechanical or chemical control methods can only provide temporary relief from conflict.

These methods parallel the techniques of frightening in overall efficiency.

There are also legal problems in the lethal control of birds. Treaties with Canada and Mexico give continual protection to migratory birds, and such birds can be destroyed only if they can be shown to cause damage. All of the order Falconiformes and such "aliens" as the house sparrow, pigeon, and starling are unprotected. By unilateral action, the U.S. Congress has given protection to bald and golden eagles. The federal government, therefore, may grant permission to kill the offending eagles and other legally protected birds if it is necessary to safeguard farm or forest crops.

By order of the Secretary of the Interior (Part 16, Title 50 of the Code of Federal Regulations), a person may destroy blackbirds, cowbirds, and grackles without a permit when they are about to commit or are committing serious depredations to ornamental and shade trees or agricultural crops. The Director of the Bureau of Sport Fisheries and Wildlife has authority to issue depredation permits to kill migratory birds that are about to cause or are causing serious damage to agricultural, horticultural, or fish-cultural interests.

If a farmer had permission to kill blackbirds, but unwittingly destroyed some legally protected mourning doves or robins, he would be in violation of the Migratory Bird Treaty Act and subject to prosecution. A game management agent of the Bureau of Sport Fisheries and Wildlife may issue individual permits to cover the accidental destruction of protected birds when a farmer engages in a campaign against blackbirds.

On private property, any campaign against unprotected birds is the responsibility of the person who suffers the damage as long as control methods present no hazard to federally protected species (Parker, 1966) or do not conflict with state laws or local ordinances. The latter can only be more restrictive than the federal regulations.

The oldest lethal techniques are shotguns and traps. Modern traps include cage, light, decoy, and elevator traps, and the electric perch (Zajanc, 1962). The effectiveness of a baited trap usually depends on competing attractants, and the light trap depends on the expertise of the user.

Control of pest birds with lethal chemicals has gained considerable momentum in the last two decades. An example is the specific avicide DRC 1339, now available commercially for starling control. However, no oral or contact toxicant has been found that is specific for a given bird, and the use of experimental toxicants presents some danger to other animal species and man. Until such time as toxicants are specific for a target species, their use should be greatly restricted.

Toxic chemicals used in bird control require federally registered labels, which give the recommendation of the manufacturer as well as the assurance of a thorough review by experts.

The introduction of predators has been an effective way of controlling herring gull populations that nest on isolated islands. In one test off the coast of Massachusetts, introduced raccoons and foxes eliminated 90 percent of the eggs and young gulls (Dykstra, 1966).

One of the untried biological controls is the use of a disease-causing organism that will be specific for a target species. Such a possibility is intriguing, but there are inherent dangers in this control method (Herman, 1964). The applicant organism must be highly pathogenic to the prospective subject species. The potential killing power, residual duration, and ultimate resistance must be anticipated. The applicant organism must be available and the host specific.

Cultural practices that make crops less available to birds will reduce damage. Deep planting often reduces loss of sprouted seed; delaying planting until after the birds migrate, or planting seeds under water, as with the rice culture in Arkansas, will also make seed unavailable; planting so that the crop matures when natural foods are abundant; and quick-ripening varieties of grain shortens the period of exposure to bird attack.

The development of damage-resistant varieties of agricultural crops is another possibility. A new variety of corn with tough, heavy husks prevents birds from easily getting at the kernels (Anon., 1952). High-tannin milo sorghum with sharp awns is another possibility (Aldrich, 1962).

The use of combines and grain driers has eliminated the period of time that crops such as corn, rice, wheat, and peanuts were susceptible to bird damage. Feeding programs on migratory bird refuges have helped to reduce damage by waterfowl to surrounding farm crops.

PREVENTIVE TECHNIQUES

Preventive techniques enabling men and birds to coexist in harmony have been given little attention until recent years.

Mechanical control efforts include the exclusion of birds from places where they cause damage, or the altering of a site to make it untenable for birds. The physical exclusion of birds and habitat alteration are suitable for small areas such as home gardens and some berry crops, but the techniques are expensive to employ on a large scale. Roosting sites for starlings, sparrows, and pigeons in cities might be eliminated by proper design of new buildings, or by alterations to existing buildings.

The biological aspect of preventive techniques for the control of bird-pest situations involves agricultural practices and some manipulation of the habitat. Alterations in time and method of planting and harvesting crops and in the variety of crops grown have been useful in dealing with agricultural problems (Buckley and Cottam, 1966). Information on bird behavior can be useful in preventing collisions with aircraft. Modification of bird habitat, such as the extensive thinning of tree-roosting sites to control starlings in towns, is a control method that offers some degree of permanence. Good housekeeping around warehouses, grain elevators, and other commercial establishments can do much to avoid bird-pest problems.

An example of the use of biological (behavioral) information occurred in preventing a bird-pest situation. An aviculturist in Joplin, Missouri, was losing prized waterfowl to great horned owls; even though he installed lights over the pond, depredation continued. He began shooting owls, but continued to lose an average of ten fowl each year. Since horned owls have a strong territorial behavior and will resist invasion by other horned owls, the aviculturist tried another scheme; he shot to scare but not to kill. The result was that the two owls occupying the territory avoided the pond and also prevented other owls from using the area. In the following two years he lost only one duck and did not have to kill any more owls.

In most cases, natural populations of vertebrate animals are regulated by density-dependent factors such as disease, predation, and lack of food that control the death rate (Lack, 1954). Density-independent factors, of which climate is a collective unit, influence density-dependent mortality. Thus, one indirect avenue for artificial control is through a manipulation of those aspects of the environment that affect the density of a given population and force it to become self-regulating. The direct approach places the controlling agent in the role of a predator or disease organism, and causes a sufficient number of animals to be removed bodily to effect control. Even so, the reproductive potential of the animal to be controlled is often so great that removal by man cannot keep abreast of recruitment.

Reproduction inhibitors have stirred the imagination of control biologists, but appreciation of their potential in alleviating bird-pest situations requires an understanding of the mechanics of animal populations. A recent development involves treated baits that chemically alter the reproductive physiology of the animal and thus eliminate progeny by preventing conception, birth, or survival. This has been referred to jokingly as "the pill for pests." If properly handled, this method could relieve a pest situation specifically, economically, humanely, and without alteration of the environment. In time, the potency and specificity of reproduction inhibitors may become refined to a point where chronic pest situations can be rapidly reduced to tolerable interactions.

In any pest situation requiring control, the action program should be one that will be most effective without seriously disturbing the ecology of other organisms. It must also be readily available and economically feasible. The kind, cost, and magnitude of control should be commensurate with the intensity and gravity of the pest impact.

APPENDIX: COMMON AND SCIENTIFIC NAMES OF BIRDS

Common Name	Scientific Name	Common Name	Scientific Name
Blackbird, red-winged	*Agelaius phoeniceus*	Gull, herring	*Larus argentatus*
Blackbird, yellow-headed	*Xanthocephalus xanthocephalus*	Heron, great blue	*Ardea herodias*
Chicken	*Gallus gallus*	Juncos	*Junco* spp.
Cowbird	*Molothrus ater*	Kingfishers	*Megaceryle* spp.
Crane, sandhill	*Grus canadensis*	Magpies	*Pica* spp.
Crow, common	*Corvus brachyrynchos*	Mallard	*Anas platyrhynchos*
Dove, mourning	*Zenaidura macroura*	Osprey	*Pandion haliaetus*
Eagle, bald	*Haliaeetus leucocephalus*	Owl, great horned	*Bubo virginianus*
Eagle, golden	*Aquila chrysaetos*	Pheasant, ring-necked	*Phasianus colchicus*
Falcon, peregrine	*Falco peregrinus*	Pigeons	*Columba livia*
Goose, snow	*Chen hyperborea*	Pintail	*Anas acuta*
Grackle, common	*Quiscalus quiscula*	Robin	*Turdus migratorius*
Grebe, horned	*Podiceps auritus*	Siskin, pine	*Spinus pinus*
Grosbeak, evening	*Hesperiphona vespertina*	Sparrow, house	*Passer domesticus*
Grosbeak, pine	*Pinicola enucleator*	Starling	*Sturnus vulgaris*
Grouse, sage	*Centrocercus urophasianus*	Swan, whistling	*Olor columbianus*
Grouse, sooty	*Dendragapus fuliginosus*	Towhees	*Pipilo* spp.
Gull, California	*Larus californicus*	Widgeon, American	*Mareca americana*

REFERENCES

Aldrich, D. J., Jr. 1962. People, pests and some plans, pp. 213–226. *In* Proc. Vert. Pest
    Control Conf. Natl. Pest Cont. Ass., Elizabeth, N.J.

Anonymous. 1952. New hybrid corn shows promise of reducing bird damage. Ohio Farm
    and Home Res. 37(279):87, 96.

Benton, A. H., and L. E. Dickinson. 1966. Wires, poles and birds, pp. 390–395. *In* Birds
    in our lives, A. Stefferud (ed). U.S. Dept. of Interior, Washington, D.C. 561 pp.

Besser, J. F. 1962. Research on agricultural bird damage control problems in western
    United States. Unpublished ms. presented at 1st Bird Control Seminar, Bowling Green
    State Univ. 4 pp.

Besser, J. F., J. W. De Grazio, and J. L. Guarino. 1968. Costs of wintering starlings and
    red-winged blackbirds at feedlots. J. Wildl. Manage. 32(1):179–180.

Bliese, J. C. 1959. Four years of battle at "blackbird roosts": a discussion of methods
    and results at Ames, Iowa. Iowa Bird Life 29(2):30–33.

Boardman, R. C. (ed.). 1967. Unlikely partners: Audubon, sheep ranchers join in eagle
    study. Audubon Leaders Conserv. Guide 8(4):2.

Buckley, J. L., and C. Cottam. 1966. An ounce of prevention, pp. 454–459. *In* Birds in
    our lives, A. Stefferud (ed.). U.S. Dept. of Interior, Washington, D.C. 561 pp.

Cappucci, D. T., Jr. 1968. Psittacosis. Nat. Comm. Disease Center, Zoonoses Surveillance,
    U.S. Dept. of Health, Education, and Welfare, Washington, D.C. 8 pp.

Craighead, J. J., and F. C. Craighead, Jr. 1956. Hawks, owls and wildlife. The Stackpole
    Co., Harrisburg, Pa. and Wildl. Manage. Inst., Washington, D.C. 443 pp.

Day, A. M. 1944. Control of waterfowl depredations. Trans. N. Amer. Wildl. Conf. 9:281–
    287.

DeGrazio, J. W. 1964. Methods of controlling blackbird damage to field corn in South
    Dakota, pp. 43–49. *In* Proc. 2nd Vert. Pest Control Conf., Univ. of California, Davis.

Dennis, J. V. 1967. Damage by golden-fronted and ladder-backed woodpeckers to fence
    posts and utility poles in south Texas. Wilson Bull. 79(1):75–88.

Dykstra, W. W. 1966. To kill a bird, pp. 447–453. *In* Birds in our lives, A. Stefferud (ed.).
    U.S. Dept. of Interior, Washington, D.C. 561 pp.

Faulkner, C. E. 1962. Urban bird problems. Unpublished ms. presented at 1st Bird Control
    Seminar, Bowling Green State Univ. 7 pp.

Frings, H. 1964. Sound in vertebrate pest control, pp. 50–56. *In* Proc. 2nd Vert. Pest
    Control Conf., Univ. of California, Davis.

Giltz, M. L. 1960. The nature and extent of bird depredations on crops. Trans. N. Amer.
    Wildl. and Nat. Resour. Conf. 25:96–99.

Hammond, M. C. 1961. Waterfowl feeding stations for controlling crop losses. Trans.
    N. Amer. Wildl. and Nat. Resour. Conf. 26:67–79.

Hayne, D. W., and H. A. Cardinell. 1949. Damage to blueberries by birds. Michigan Agric.
    Exp. Sta. Quart. Bull. 32(2):213–219.

Herman, C. 1964. Disease as a factor in bird control, pp. 112–119. *In* Proc. 2nd Bird
    Control Seminar, Bowling Green State Univ.

Herman, C. M. 1966. Birds and our health, pp. 284–291. *In* Birds in our lives, A. Stef-
    ferud (ed.). U.S. Dept. of Interior, Washington, D.C. 561 pp.

Hickey, Joseph J. (ed.). 1969. Peregrine falcon populations/their biology and decline.
    The Univ. of Wisconsin Press, Madison, Milwaukee, and London. 596 pp.

Horn, E. E. 1949. Waterfowl damage to agricultural crops and its control. Trans. N. Amer.
    Wildl. Conf. 14:577–586.

Horn, E. E., and L. L. Glasgow. 1964. Rice and waterfowl, pp. 435–443. *In* Waterfowl tomorrow, J. P. Linduska (ed.). U.S. Dept. of Interior, Washington, D.C. 770 pp.

Kottman, R. M. 1967. Irresistible force—immovable object, pp. 5–9. *In* Proc. N. Amer. Conf. on Blackbird Depredation in Agriculture. Ohio State Univ., Columbus. 62 pp.

Lack, D. 1954. The natural regulation of animal numbers. Clarendon Press, Oxford. 343 pp.

Lagler, K. F. 1939. The control of fish predators at hatcheries and rearing stations. J. Wildl. Manage. 3(3):169–179.

Lawrence, W. H., N. B. Kverno, and H. D. Hartwell. 1961. Guide to wildlife feeding injuries on conifers in the Pacific Northwest. West. Forest. and Conserv. Ass., Portland, Ore. 44 pp.

Marsh, G. A. 1962. Bird borne disease. Unpublished ms. presented at 1st Bird Control Seminar, Bowling Green State Univ. 2 pp.

Martin, M. D. 1942. Wyoming magpie control. Wyo. Wild Life 7(10):13–15.

McAtee, W. L. 1911. Woodpeckers in relation to trees and wood products. U.S. Dept. of Agriculture Biol. Surv. Bull. 39. Washington, D.C. 99 pp.

Meinertzhagen, R. 1959. Pirates and predators. Oliver and Boyd, London. 230 pp.

Miller, C. E. 1967. A congressional view of the blackbird problem, pp. 56–60. *In* Proc. N. Amer. Conf. on Blackbird Depredation in Agriculture. Ohio State Univ., Columbus. 62 pp.

Mitchell, R. T. 1960. Management to avoid bird depredations. Trans. N. Amer. Wildl. and Nat. Resour. Conf. 25:99–101.

Parker, L. A. 1966. The legal basis, pp. 433–437. *In* Birds in our lives, A. Stefferud (ed.). U.S. Dept. Interior, Washington, D.C. 561 pp.

Paynter, E. L. 1966. Crop insurance. The Sask. Govt. Ins. Office, Regina, Sask. Mimeo. 4 pp.

Price, E. R. 1966. Histoplasmosis—its relationship to birds, pp. 1–8. *In* Proc. 3rd Bird Control Seminar, Bowling Green State Univ.

Quinn, I. T. 1952. Winter feeding of wild geese on wheat fields helps crops. Va. Wildl. 13(12):12.

Seubert, J. L. 1965. Biological studies of the problem of bird hazard to aircraft. Final Report, project agreement FAA/BRD A-90, report no. RD-65-73. U.S. Dept. of Interior, Bureau Sport Fisheries and Wildlife. Washington, D.C. 22 pp.

Seubert, J. L. 1966. Biological control of birds in airport environments. Interim report, project agreement FA65WAI-77, project no. 430-011-01E, SRDS Report no. RO-66-8, U.S. Dept. of Interior, Bureau Sport Fisheries and Wildlife, Washington, D.C. 41 pp.

Seubert, J. L. 1967. What's being done about blackbird control at the Bureau of Sport Fisheries and Wildlife Research Center Eastern Branches, pp. 42–45. *In* Proc. N. Amer. Conf. on Blackbird Depredation in Agriculture, Ohio State Univ., Columbus, 62 pp.

Seubert, John L. 1968. Control of birds on and around airports. U.S. Dept. of Interior, Bureau of Sport Fisheries and Wildlife, Division of Wildlife Research, Washington, D.C. FAA Rpt. no. RD-68-62. 30 pp.

Siebe, C. C. 1964. Starlings in California, pp. 40–42. *In* Proc. 2nd Vert. Pest Control Conf., University of California, Davis.

Smith, C. F., and S. E. Aldous. 1947. The influence of mammals and birds in retarding artificial and natural reseeding of coniferous forests in the United States. J. Forestry 45(5):361–369.

Smith, R. N. 1966. Control of birds in aircraft hangars, pp. 30–33. *In* Proc. 3rd Bird Control Seminar, Bowling Green State Univ.

Solman, V. E. F. 1966. Ecological control of bird hazard to aircraft, pp. 38–56. *In* Proc. 3rd Bird Control Seminar, Bowling Green State Univ.

Spear, P. J. 1966. Bird control methods and devices—comments of the National Pest Control Association, pp. 134–143. *In* Proc. 3rd Bird Control Seminar, Bowling Green State Univ.

U.S. Air Force. 1966. Bird/aircraft collisions. Air Force Office of Scientific Research, Office of Aerospace Research, USAF, Arlington, Va.

U.S. House of Representatives. 1966. Predatory mammals. Hearings before the Subcommittee on Fisheries and Wildlife Conservation of the Committee on Merchant Marine and Fisheries. 89th Congress, 2nd session, serial 89-22. U.S. Govt. Printing Office, Washington, D.C.

U.S. Senate. 1962. Protection for the golden eagle. Hearing before a subcommittee of the Committee on Commerce. 87th Congress, 2nd session on S.J. Res. 105 and H.J. Res. 489, June 26, 1962. U.S. Govt. Printing Office, Washington, D.C.

Wagar, J. V. K. 1946. Colorado's duck-damage, grain-crop problem. Trans. N. Amer. Wildl. Conf. 11:156–162.

Williams, F. J. 1960. The agricultural experiment station in relation to bird depredations. Trans. N. Amer. Wildl. and Nat. Resour. Conf. 25:113–116.

Williams, C. S., and J. A. Neff. 1966. Scaring makes a difference, pp. 438–445. *In* Birds in our lives, A. Stefferud (ed.). U.S. Dept. of Interior, Washington, D.C. 561 pp.

Zajanc, A. 1962. Methods of controlling starlings and blackbirds, pp. 190–212. *In* Proc. Vert. Pest Control Conf., Nat. Pest Control Ass., Elizabeth, N.J.

CHAPTER 5

Small-Mammal Pest Situations

Small mammals, including rodents, hares, rabbits, bats, and some carnivores, are involved in a great variety of pest situations. Of these, rodents are most frequently associated with problems. The order Rodentia comprises about half of all mammalian species because they are so highly productive and widespread.

Control procedures used to alleviate problems associated with small mammals are as varied as the pest situations. In general, population reduction provides temporary relief, whereas environmental control can produce more lasting effects. In situations such as the destruction of tree seed by rodents, temporary protection is all that is required. In urban rat problems, the opposite is true: the ultimate solution must be removal or destruction of rat food and harborage.

Prerequisite to a solution is an understanding of the specific problem. Biological information concerning the animal species involved, an understanding of the conditions of the problem, and knowledge of the control procedure (including possible hazardous side effects) are essential. Conflicts between man and small mammals are continual, with each situation presenting some unique aspects. The infinite variety precludes a detailed account, but principles of control will be considered under four categories: agriculture, industry, recreation, and health.

AGRICULTURE

As man manipulates and modifies his environment to produce food, fiber, and other products essential to civilization, he encounters problems associated with the activities of naturally existing small-mammal populations. These

83

problems arise in a wide range of situations: in orchards, croplands, rangelands, or forests. Most often the mammals involved are rodents of various kinds that destroy agricultural products man intends for his own use, but problems also may arise with other herbivores and with small predators.

Rodent populations often are affected directly by man's activities, as when he improves their living conditions by planting favored foods and cover crops or creates preferred rodent habitats by cutting forests or permitting overgrazing on rangelands. In such situations, rodent populations often increase rapidly to problem levels. On the other hand, small mammals may be nearly eliminated over vast areas of cultivated land by repeated mechanical disturbance of their habitats, as in corn or wheat fields.

PROBLEMS IN ORCHARDS

Fruit trees may be damaged by several types of small mammals, including mice, rabbits or hares, pocket gophers, tree squirrels, ground squirrels, and others. The most serious and extensive damage results from bark-eating by mice in grassy orchards. These are usually microtine species that find an ideal habitat in grassy ground cover. Clean cultivation usually eliminates them, but in many orchards, especially apple orchards, a sod cover is maintained. Under such conditions, meadow mice and pine mice in the United States may destroy or weaken orchard trees of all ages. In the absence of control measures, serious losses may result.

Two types of mouse damage, one above and the other below ground, are especially prevalent in the United States. Meadow mice of the genus *Microtus* remove bark from the base or root crown of trees and may kill trees by girdling (ringing) them completely. Pine mice of the subgenus *Pitymys* are subterranean and affect trees severely by eating the small roots and debarking main roots.

Control

Various kinds of reductional and preventive methods are used to control small mammal damage in orchards. Most common are the use of poison baits, physical barriers, and habitat modifications. For example, meadow mice in orchards are usually exposed to poison baits at the end of the tree growing season, because bark feeding is most prevalent in the dormant months that follow. The bait may be poisoned grain, broadcast in various ways, or pieces of poisoned fresh fruit placed by hand in runways. A special machine may be used to create artificial subsurface tunnels in which poisoned baits are placed automatically

at intervals (Ward and Hansen, 1960). Airplanes have been used widely to broadcast grain baits efficiently and economically.

Along with programs to reduce the animal populations, many fruit growers
place physical barriers of wire mesh or other material around the bases of
younger trees. Habitat control, consisting of frequent mowing to reduce mouse
food and shelter, may be used also; or vegetation may be eliminated near the
trees with chemical weed killers or by mechanical means.

Underground feeders, such as pine mice, are more difficult to control, and
hand-placed baits are usually necessary. Some of the more toxic insecticides
are effective as ground sprays for killing mice (Horsfall, 1953; Eadie, 1959),
including pine mice, but the advisability of using them is questionable because
of indiscriminate, long-range, and widespread effects on other animals.

Repellent chemical sprays on trees are not used often against mice because
of the difficulty of coating root systems, but such materials may offer protection against hare and rabbit damage.

Assessment

Orchard problems with small mammals may be solved effectively and economically by current techniques, but since highly toxic poisons may cause secondary poisoning of nontarget animals, narrow-spectrum rodenticides should be
developed. Habitat manipulations to minimize mouse populations are not always used as effectively and consistently as they might be.

PROBLEMS IN CULTIVATED CROPS

Many kinds of small mammals, especially rodents, damage cultivated crops
(Table 1). Among these are various native rats and mice, squirrels, woodchucks,
pocket gophers, and hares and rabbits.

Whenever ground squirrels occur near agricultural crops, they may attack
grains, vegetables, fruits, and grasses. The most extensive damage is to grain
and pasturage. A study of one ground squirrel species revealed that each squirrel was capable of destroying over 50 pounds of wheat per season (Shaw, 1920).
A population of 12 squirrels per acre was found to decrease the annual dry
forage yield by more than 1,000 pounds per acre, or by about 38 percent, in
enclosures (Fitch, 1948).

Marmots or woodchucks may also occur in pest situations near crops. Individuals may eat a pound of green food daily and destroy more by trampling in
their runways. Their soil mounds may interfere with mechanical harvesting.

Meadow mice numbers fluctuate greatly. When abundant, their depreda-

TABLE 1 Some Problem Mammals of North America in Relation to Orchards and Cultivated Crops

Animal	Problem	Current Controls
Meadow mice	Remove bark in orchard trees; destroy vegetable and forage crops	Poison baits, habitat control, traps, mechanical and chemical repellents
Pine mice	Eat roots and bark from orchard trees; damage root crops, bulbs, and seeds	Poison baits, habitat control, ground sprays
Muskrats	Damage gardens, pond dikes, and irrigation ditches	Traps, poison baits, dike barriers
Cotton rats	Destroy sugar cane, cotton, truck crops, grain, and planted seeds	Poison baits, seed repellents
Rice rats	Attack vegetable crops, cotton, rice, fruit trees	Poison baits, habitat control
White-footed mice	Destroy planted seeds, corn in shocks, standing grain in field borders	Poison baits, seed repellents, modified farming practices
House rats and house mice	Destroy and contaminate stored fruits, vegetables, and grain; damage sugar cane and other field crops	Rat-proofing storage areas, poison baits, traps, fumigation
Kangaroo rats and pocket mice	Eat range forage and seeds; damage contour dikes in irrigation	Poison baits, modified range management, seed repellents
Ground squirrels and chipmunks	Damage or destroy vegetables, grain, forage plants, nut and fruit trees, seeds, bulbs	Poison baits, fumigation, traps, habitat manipulation, repellents
Woodchucks (marmots)	Destroy gardens and field crops; burrows and soil mounds affect harvesting and irrigation structures	Poison baits, fumigation, trapping, shooting
Tree squirrels	Damage garden crops, nuts, and fruits	Trapping, shooting, mechanical repellents
Pocket gophers	Destroy root crops, range forage; damage orchard trees, underground cables, irrigation structures	Poison baits, traps, habitat manipulation, herbicides
Cottontail rabbits	Damage fruit trees, gardens, tree nursery stock	Repellents, fences and tree guards, trapping, shooting
Hares and jackrabbits	Damage fruit trees, range forage, garden crops	Repellents, poison baits, shooting, fences, and guards, trapping
Raccoons	Destroy garden crops, especially corn	Shooting, trapping, electric fence

Source: Eadie, W. R. 1954. Animal Control in Field, Farm, and Forest. Macmillan Co., New York.

tions to crops may sometimes reach disastrous proportions (Vertrees and Spencer, 1959). In lesser numbers they take a steady toll of vegetation, which, in the aggregate, can be significant. Bailey (1924) estimated that as few as 10 mice per acre (a very low population) would consume 11 tons of grass or alfalfa per 100 acres each year (about 5 percent). Recent evidence indicates that 100 mice per acre can consume 4 percent of the annual production in alfalfa (Rudd, 1964). Meadow mice may also destroy vegetables in truck gardens to a serious degree.

White-footed mice, cotton rats, rice rats, pocket mice, and kangaroo rats may at times be destructive in croplands.

Pocket gophers are serious in agricultural pest situations and destroy many kinds of crops. In addition to feeding on roots and surface vegetation, these tunnel-diggers may bury significant portions of the crop with excavated soil. They have been found to destroy from one-fourth to one-third of the alfalfa crop over large areas (Miller, 1953). Their burrows may cause costly breaks in irrigation ditches.

Hares and rabbits are frequent pests in gardens and croplands. In irrigated regions jackrabbits may seriously damage all kinds of crops.

Control

Most methods used to control damage in croplands are aimed at destroying the offending small mammals. Poisoning, trapping, shooting, and fumigation are among the methods used. Of these, poisoning is most widely used and probably the most economical and effective.

The poison used depends upon acceptability and effectiveness for the target animal. For example, strychnine and sodium fluoroacetate (1080) are used against ground squirrels but not ordinarily against meadow mice, for which zinc phosphide is usually preferred.

Grain is frequently employed as a bait, because poisoned grain is easily stored and adapted to mechanical distribution methods. Fresh fruits and vegetables also are used to carry the poison in some situations. The season of the year, availability of other foods, and acceptability to target and nontarget animals are considerations in the choice of carrier.

Ground-spray formulations of some of the more toxic insecticides have been used in some types of crops (e.g., seed-crop alfalfa) to poison meadow mice.

Traps appropriate to the target animal are sometimes useful on small areas, but are not often feasible on large areas. Shooting may be effective against the larger rodents (e.g., woodchucks) or rabbits and hares on limited areas, but not against smaller species. Toxic gases have some utility in fumigating limited burrow systems, such as those of woodchucks and ground squirrels.

Repellents for protecting field crops from small mammal attacks are seldom economical on large areas, and such chemicals, if effective, are not easily removed from edible plant parts and may be toxic to man or livestock. The most useful repellents available are preparations containing fungicides such as zinc dimethyl dithiocarbamate, or tetramethyl thiuram disulfide (Besser and Welch, 1959). These, formulated with efficient adhesives to resist weathering, are useful on nonedible plant parts on small areas where cost factors are not critical.

Physical barriers or fences, electrified or not, may have limited utility in special situations, but are often ineffective because the animals are able to circumvent them. Even where successful, costs and inconvenience may preclude their use on large acreages.

Assessment

The farmer who suffers measurable crop losses to rodents or other small mammals is likely to apply direct measures for destroying the pest species. Because this commonly involves use of poison baits, one current need is to develop more precisely tailored lethal agents that will kill only the target species; or, at least, fewer nontarget species than are affected by toxic chemicals now in use.

Selective placement of poison baits may aid in restricting their availability to target animals, as when bait is placed in ground squirrel burrows or pocket gopher tunnels.

PROBLEMS IN FOREST CROPS

All animal life is dependent on plants for its existence: forest wildlife is no exception. The removal of seeds and foliage of "undesirable" plants by wildlife often helps the land manager; "damage" is involved only when the losses or injuries conflict with man's interests.

Unfortunately, the feeding activities of wildlife and the interests of forest land managers are often at variance. The degree of incompatibility generally depends on how intensively the land is managed. In the past, with seemingly unlimited supplies of old-growth timber, there was little concern about animal damage to natural forest reproduction. However, under sustained yield management, forest regeneration became paramount, resulting in increased awareness of wildlife damage.

Forests in all regions are subject to wildlife injuries; only the type and severity differ. The three broad categories of damage are seed removal, clipping and browsing, and bark and root injuries.

The loss of tree seed is the least apparent form of wildlife damage, but one of the most serious problems. Millions of acres of artificially and naturally seeded forest lands have failed to regenerate because the seeds were eaten by various species of mice, chipmunks, squirrels, birds, and even shrews. Seed caching by rodents can be equally devastating. Many of these animals damage newly emerged seedlings also. Considering that deer mice (*Peromyscus* spp.) inhabit most areas of the North American continent, and that each mouse weighing less than one ounce can consume more than 200 Douglas-fir (*Pseudotsuga menziesii*) seeds in one night, the potential of these losses becomes apparent when the rate of artificial seeding is about 20,000 seeds per acre.

Clipping and browsing damage to established seedlings, another major concern, involves rodents (pocket gophers and porcupines) and lagomorphs (rabbits and hares). Clipping or cutting often causes high rates of seedling mortality, particularly in new plantations, where seedlings are frequently cut at or near ground level by hares or rabbits. The severity of this problem was exemplified in a study in which a single snowshoe hare was placed in a vegetated one-acre enclosure planted with two-year-old Douglas-fir seedlings. Of 500 seedlings present, 478 were clipped by the hare during the first week. Spared from damage, most plantations will outgrow their susceptibility to clipping within a few years. In contrast, the browsing of seedlings is seldom a major mortality factor, but often is responsible for retarding plantations for long periods.

Bark injuries are the most varied in terms of animals and age of trees involved. Meadow mice (*Microtus* spp.) girdle seedlings, whereas bears are responsible for similar damage to large trees. Porcupines remove bark from trees of all sizes. Other animals that inflict bark injuries are mountain beaver (*Aplodontia rufa*), rabbits, hares, squirrels, and pocket gophers. Root injuries are most frequent on seedlings or small saplings, and pocket gophers are usually the animals responsible. Root cutting in plantations is seldom apparent until crowns turn brown during summer drought, or seemingly healthy trees are tipped to odd angles by wind.

Control

The scope and complexities of forest wildlife damage preclude simple solutions. The animals involved often are game species or cohabitant with game species. The dynamics of some rodent populations defy efforts to keep their numbers at desirable levels. These problems are compounded by their duration in the regeneration cycle of the trees; animal damage starts with the seed, or even the cone bud, and continues for several decades after establishment.

In general, there are two approaches to reducing forest wildlife damage: reduce the number of animals, or make the plants less attractive or available. Population reduction can either be direct, through lethal agents or regulated harvest by hunters; or indirect, through habitat manipulation or limiting reproduction. Attractiveness or avilability can be reduced by mechanical or chemical barriers, by altering the internal composition of the plant, or by replacing the plant species with one less susceptible to damage.

Available methods of protecting seed from rodents include coating the seed with an endrin formulation, using lethal agents to reduce rodent numbers, and placing seed inside mechanical barriers (Kverno, 1964). The endrin treatment is most effective in the Pacific Northwest. In the pine regions, baiting with rodenticides may successfully complement the endrin treatment. In the East, tin cans are being used as mechanical barriers to help protect walnuts and acorns from squirrels. Application of contact repellents on foliage is the most effective known method of reducing rabbit and hare damage to seedlings. Lethal agents applied to fruit, vegetable, or grain carriers will reduce populations of small rodents, and specially prepared wooden blocks containing salt and strychnine are effective in controlling porcupines.

Assessment

In forestry, the transition from resource utilization to resource management is relatively recent. Consequently, only interim methods of alleviating wildlife problems are now available. Greater sophistication in methods, materials, and techniques of application are prerequisite to integrated management.

INDUSTRY

PROBLEMS WITH COMMENSAL RODENTS

The economic problems associated with small mammals in industry relate primarily to commensal rodents. The term *commensal* means "eating at the same table." In the United States this group includes Norway rats, three subspecies of black rats, and house mice.

In the interior and northern part of the country, only Norway rats and house mice occur. In the coastal regions, particularly in the South, the black rats often predominate. Frequently more than one form is found in the same area, and because of their somewhat different habits and food preferences, the control methods have to be modified accordingly.

Accurate estimates of nationwide damage by commensal rodents are not available. In contrast to the systematic appraisals made for other wildlife populations, estimates of numbers of commensal rodents and the damage they do have been inadequate. National or regional computations commonly are merely expansions of local estimates. Even without accurate figures, it is generally conceded that commensal rodents are foremost among small mammals in causing economic losses.

Industrial losses resulting from the activity of commensal rodents are as varied as the situations in which they are found, but are of three general types: consumption of foodstuff, contamination of foodstuff, and gnawing damage to miscellaneous objects.

Consumption of food products by rodents is a problem in nearly all phases of food processing. It is particularly serious in milling and storing cereal grains and products. Of more concern than the quantity of material consumed is the contamination of even greater amounts with urine, fecal matter, hair, and other materials.

Gnawing is essential to the welfare of rodents, because the four incisor teeth grow continually and require constant use to maintain them at normal length. In their quest for food, rodents will gnaw through almost any material offering an edge that is less hard than their dental enamel. They readily penetrate corrugated boxes, wooden doors, and occasionally even ratproof concrete before it has hardened completely. In addition to gnawing to gain entry, rodents will gnaw such items as furniture, lead pipes, plastic products, and electrical wiring.

PROBLEMS WITH FIELD RODENTS

There are many other small mammals that contribute to the animal damage sustained by industry. One increasing concern is gnawing damage to cables and wires by field rodents. Buried telephone cable is frequently cut by pocket gophers, and power failures sometimes result when tree squirrels gnaw aerial cables. Of greater significance because of its frequency is cutting of surface cables by rabbits, hares, wood rats, ground squirrels, and several species of mice. Seismograph companies, the Atomic Energy Commission at its test sites in Nevada, and even National Aeronautics and Space Administration projects have been plagued by rodent damage to cables. Replacement of the cables is costly, but the program delays are even more serious.

The military forces have experienced difficulties throughout the world with gnawing damage to communication systems. Recently, rodent damage to electrical wiring in helicopters has been reported in Vietnam. In such situations,

where national defense or human lives are involved, the economic loss becomes
insignificant compared with other values.

Control

Procedures for rat and mouse control are oriented toward preventive measures
and reductional control. Each of these approaches is dependent on the physical and economic conditions prevailing in a given situation. In areas of relatively
modern construction, preventive measures cost less and provide more permanent protection than reductional control. Conversely, in areas of old construction, ratproofing is often not practical, and reductional control is the most
logical approach. A combination of both approaches in conjunction with a
sanitation program may be required for satisfactory results.

Understanding the behavior of the animals involved is essential to intelli-
gent application of control methods. This principle applies particularly to control of commensal rats. They frequently establish habitual movement patterns
and feeding periods that become almost ritualistic, but they are also extremely
wary of new items in their environment. Baits are often rejected when newly
encountered; even placebo baits are poorly accepted at first. "Bait shyness" is
best remedied by conditioning the rats to the untreated carrier for several days
before introducing treated material.

Knowledge of the species composition of an offending rodent population
helps in selecting the control procedures. For example, Norway rats are prone
to burrowing, and are generally found at the ground level; but the black rats,
having a propensity for climbing, are associated with upper floors and attics of
a building.

There has been more progress in the past two decades in developing chemical agents for control of commensal rodents than during the preceding two
centuries. Foremost among recent materials are the anticoagulants that cause
bleeding and death. Without sacrificing effectiveness, they offer a degree of
safety to man and domestic animals heretofore unavailable with lethal agents.
Another recently registered material claiming rat specificity is a vasoconstrictor
that causes death by impairing circulation (Crabtree, Robison, and Perry, 1964).

Regardless of the control agent used, complete success is seldom achieved.
To prolong low population levels, a second campaign using a different bait and
lethal agent is often necessary.

Preventing rodent entry into structures is an effective method of minimiz-
ing damage. The cost of rodentproofing existing buildings can be prohibitive,
but for new construction it is less expensive and, in some municipalities, man-
datory. Ultrasonic (high-frequency) sound repellents do not obviate the necessity for ratproofing.

Chemical repellents for protecting packaged food products are effective for only a few weeks at most. After being initially repelled, the rodents use their protruding incisors to cut through the treated cartons, making minimal contact with the chemical. Contamination hazards preclude use of many chemicals known to provide greater protection. Multiwalled polyethylene tarps with an internal chemical layer permit use of compounds having a higher degree of activity but are effective for only short periods. A thin layer of stainless steel, sometimes incorporated in corrugated paper cartons, was believed to be the only permanently effective ratproof packaging, until recent reports of the effectiveness of wood and laminated paperboard containers impregnated with alums and silicates (Anon., 1968). These chemicals impart a hardness that resists gnawing by rats.

To achieve the degree of control desired, it is often advisable to enlist the support of the entire community. A systematic campaign incorporating reductional control, preventive measures, and provision for periodic inspection is advantageous. Intimately connected with rodent suppression and ratproofing is the simple matter of good housekeeping. Removal of the animal's food supply and reduction of available cover are major deterrents to reestablishment of rodent populations. Good sanitation practices are also essential to effective, long-range rodent control.

The difficulty of protecting cables from rodents is similar to that encountered with packaged materials: both are barrier-type problems, i.e., the treated material is gnawed and discarded, not eaten. Recent research has produced a rodent repellent cable coating for polyethylene and other plastic-jacketed wire and cable. In 19 months of field testing, rodent damage on an Army communications wire was reduced 92 percent when treated with an organic tin compound. For pocket gophers, however, armored cable is now the only sure protection.

Assessment

Rodent control in the United States is carried out by numerous private groups and public agencies. With commensal rodents, economic loss and health problems are so interrelated that separation is impractical and often merely academic. Contamination of food products can be a serious health menace when not detected, and a serious economic loss when discovered. Because of this close relationship, responsibilities are not always clearly defined and frequently overlap. Needless to say, the health implications predominate; therefore, health authorities are generally responsible for long-range control and community planning. In many cities, control is entirely a function of local authorities. State health departments are instrumental in enforcing state laws governing these matters, and they cooperate with city and country officials. Similarly, the U.S. Public Health Service has responsibilities in safeguarding health on the national level and cooperates with the states in this matter.

When officially supervised campaigns are not in operation or are not applicable to a situation, control on private property is often undertaken by the owner or a hired pest control operator. Generally, the means for implementing control is available, but coordination of effort is lacking, especially on a nationwide basis. Regulations governing control procedures and improvements in sanitary standards are needed. Municipal ordinances should be extended to ensure that new construction and repair conform to the best principles of rodent exclusion.

Problems with field and commensal rodents are national in scope, and the materials, methods, and other requirements for dealing with them are similar. Areas requiring expanded research include development of more effective repellents, and safer and more specific methods of reductional control. There is also a need for more knowledge of the physiology and behavior of rodents. Greater understanding of this important and dynamic animal group will better equip man for managing them.

RECREATION

In justifying control of animal pest problems, it is usually acknowledged that the designation of a "problem" as such must be related invariably and solely to man's interests. It is also common to balance a problem species' assets against its liabilities before deciding on a solution. When dealing with animal pest problems in relation to human recreational pursuits, a new dilemma is encountered: the source of the problem may often be, in part, the object of the pursuit.

On the positive side, small mammals play an important part in outdoor recreation; they are hunted, trapped, photographed, tamed, or merely observed and enjoyed in their natural state. They also add to human recreation indirectly in their roles as soil tillers, fish-habitat creators, erosion and flood controllers, insect eaters, or prey and buffer species for other desirable animals. But they may also detract from human recreation by spreading diseases, damaging cabins and other facilities or materials, eating or contaminating stored foodstuffs, preying upon game and other desirable animals, creating fire hazards, draining fish ponds, or defacing lawns, parks, gardens, and golf courses.

Direct conflicts between small mammal activities and human recreational interests are usually localized and of minor importance compared with other problems with small mammals, and they receive correspondingly less public

attention. Some outdoor enthusiasts probably regard recreational problems
created by small mammals as nuisances that can be alleviated—when necessary
or worth the effort—by common sense. But nonchalance fades when a horse
breaks its leg stepping into a woodchuck hole; when a prized dog is badly injured by porcupine quills; when a cabin burns to the ground because of rodents' gnawing matches; or when a child dies of tick fever after handling a
chipmunk!

As technological and economic progress continues to increase leisure time,
the demand for outdoor recreation will take more people to more places more
often, multiplying the opportunity for conflicts with small mammals. The
local problems thus created may or may not change, but their aggregate importance certainly will grow.

DAMAGE TO FACILITIES

Among the more important economic losses is damage by small mammals to
recreational facilities, including cabins and their contents, outbuildings, signs,
picnic tables, and wooden-handled tools. Porcupines, attracted by salt-
impregnated wood, are major offenders, but bats, woodchucks, ground
squirrels, chipmunks, wood rats, Norway rats, house mice, and some native
mice also will enter buildings and damage items such as bedding, furniture,
clothing, and foodstuffs. Bats are mainly unwelcome because of odors from
their attic roosts, their distracting noises, and their role as disease hosts. Rats
and mice often are, or become, commensal with man, invading even occupied
buildings to wreak damage.

Burrowing rodents may drain fish ponds, block ditches, and undermine
ditchbanks and roadways, some of which are used in recreation. The beaver,
highly regarded for regulating stream flow and providing trout habitat, may
also cause siltation, flood access roads, impede fish spawning, and detract
from fish habitat by cutting shade trees along streams. These losses are not
readily calculated in terms of dollars and cents, but their total significance
should not be discounted. In contrast, moles, pocket gophers, and mice often
play havoc with ornamental plantings, lawns, gardens, and golf greens. Damage
by these animals is readily visible and may be costly to remedy.

Most of the foregoing examples involve aesthetic as well as economic de-
preciation. Other, more subtle depredations cannot be appraised, but constitute an aesthetic loss. Among these are small-mammal predation on song and
game bird nests, destruction of game food and cover, suppression of forest
regeneration, and disfigurement of recreational areas. Moreover, values are
entirely relative where aesthetics are concerned. Evidence of gopher activity

is of interest to nature lovers, but a picnic ground pocked with gopher mounds is an eyesore. A woodrat nest is perfect tinder for a carelessly thrown match, but its slight hazard is offset by its value to human observers.

Control

Because of the aesthetic values involved, wholesale campaigns to eliminate small mammals will seldom be indicated in recreation areas, especially those in nonurban situations. The solution should be as localized as the problem, and as specific for the offending species or individual as possible. Limited reductional control of small mammals usually provides only a temporary solution because of rapid population recovery and reinvasion. Therefore, when possible, control should be oriented toward protection of the facility rather than elimination of the animal. This can be achieved with physical barriers, chemical repellents, or habitat modifications.

Cabins and other buildings can be rodent-proofed by sealing all inlets and screening drainpipes, vents, and crevices. Destruction of the offending species' habitat and food supplies in the immediate vicinity is also advantageous. Tight-fitting garbage can lids and removal of nearby cover help. In some cases, protection can be provided by application of chemical repellents to plants and structures. Chemicals and techniques to be employed vary with the situation and species involved.

As an example, moles, which damage golf greens and lawns by excessive tunneling, may be controlled in such areas by chemical treatments that eliminate worms and insect larvae, which are the chief food of the moles. On small areas, special mole traps may be used to remove the animals themselves.

Regardless of other measures taken, encouragement of native predators is always a useful supplement. Predation seldom eliminates a small mammal population, but often reduces it, thus lessening population pressures that might encourage invasion of man-made facilities.

When and where game regulations permit, public hunting of suitable problem animal populations should be encouraged. Like predation, sport hunting is seldom a complete solution, but it can serve to control damage and at the same time provide recreation. Hunters naturally seek out areas of high animal density and avoid areas that afford little shooting. Thus, public aid in control should naturally tend to be applied where needed most, without destroying the populations. Some species that may present damage problems but are not ordinarily thought of as edible game are, in fact, quite palatable: among these are marmots, porcupines, wood rats, and prairie dogs. If this information were publicized, perhaps greater public participation in animal damage control could be expected.

HUMAN HEALTH AND SAFETY

Many serious diseases are transmitted to man directly or indirectly by wild or commensal small mammals. These are treated in more detail elsewhere in this report, but should be mentioned here in relation to recreation. Again, hazards to human health through recreational contact with small mammals are usually isolated in time and space. There is no reason to expect the problem to reach epidemic proportions even though recreational activities continued to engage more people more often, but the number of incidents will surely increase unless preventive measures are taken. The recent vogue for camping, wild-river canoeing, and other outdoor recreation is particularly important in this respect, since man will come into direct and indirect contact (through pets, contaminated objects, and foodstuffs) with wild species. All small mammals are potential disease carriers or transmitters, but of special concern are carriers of bubonic plague, rabies, tularemia, typhus, and tick fever.

Normally, few rodents are either inclined or able to inflict physical injuries on man, but most will struggle and bite or scratch in self-defense. Aside from the danger of infection or disease transmission, such wounds are rarely serious. Household pets encountering porcupines or other large rodents, especially for the first time, may be injured, but seldom fatally. As a result of their burrowing activities, certain small mammals may indirectly cause injuries to human beings, horses, or big game.

Capturing and selling wild rodents (such as ground squirrels) as pets is a growing industry in some states. Despite legally imposed quarantine periods, diseased or disease-carrying animals have been sold, with consequent disastrous results. Closer supervision or more detailed regulation may be needed if this health hazard is to be minimized.

Control

Because small mammal problems in a recreational context sometimes involve private ownership, the responsibility for control may be assumed by the individuals concerned. This in itself constitutes a hazard to human health and safety, since highly toxic materials and traps capable of injuring pets and unwary people may be used carelessly. Government-sponsored public education via schools, posters, leaflets, and mass-communication media may be an effective means of limiting these hazards.

In guarding human health, the rule should be to avoid contact with small mammals and their parasites where risk is involved. Prolonged or frequent recreational activity in areas where certain animal-borne diseases are known or suspected to be endemic calls for preventive inoculations. Use of contact repellents, applied to skin and clothing, against ectoparasitic vectors should be encouraged. Careful attention to sanitation is advisable, particularly where there is danger of contamination of foodstuffs, clothing, and bedding by disease-transmitting rodents.

When necessary, small mammal problems may be alleviated by shooting or by judicious use of traps and poisons. Because of the extra hazards to children, pets, and desirable wild animals in areas used for recreation, expert advice should be sought before using such drastic measures.

Assessment

Small mammal problems connected with human recreation are almost invariably local, seldom serious, and therefore not well-documented on a national basis. The fact that large-scale appraisal of the problem has not been undertaken may reflect its relative unimportance. The Interior Department's recently created Bureau of Outdoor Recreation in cooperation with the U.S. Public Health Service might well devise means of assessing the problems so that public agencies can advise or aid in control.

Continuing ecological and damage-control research will bring solutions to many animal problems, and techniques for application in recreation areas will surely evolve as by-products. For example, weather-resistant formulations of stable chemical repellents developed to protect forest plantations might prove to be useful in guarding cabins and other recreational facilities against gnawing damage. Development of monospecific toxic agents that present less hazard to human beings, pets, or predators would also be valuable.

HEALTH

Small mammals may be vectors for a variety of diseases transmissible to man, domestic animals, or livestock. Commensal rats and mice are involved most often, but many native species may also transmit diseases (Table 2).

PROBLEMS WITH RODENTS, HARES, AND RABBITS

Because of their close association with man, commensal rats and mice have long been implicated in the transmission of serious diseases to man and his domestic animals. Historically, bubonic plague and typhus have been the most serious and dramatic of the many diseases rats may transmit to man, and millions of people died in vast epidemics. With present control methods, serious outbreaks of plague are confined mostly to tropical climates and less-developed

TABLE 2 Some Problem Mammals of North America in Relation to Health

Animal	Problem	Current Controls
House rats and mice	Transmit plague, typhus, leptospirosis, rat-bite fever, *Salmonella*, etc.	Anticoagulant poisons, other poisons, ratproofing, sanitation, fumigation
Wild rodents	May transmit spotted fever, plague, typhus, tularemia and other diseases	Trapping, poison baits, skin and clothing repellents, immunization, avoidance of contact
Rabbits and hares	Transmit tularemia, plague, spotted fever	Shooting, poison baits, avoidance of contact, personal repellents, immunization
Carnivores	Transmit rabies	Immunization of dogs, humans in risk areas; wild population reductions by trapping, poisoning
Bats	May transmit rabies, histoplasmosis, Chagas' disease, and others	Exclusion, avoidance of contact, fumigation, contact poisons, netting

Source: Eadie, W. R. 1964. Animal Control in Field, Farm, and Forest. Macmillan Co., New York.

countries such as India, where thousands still die from the disease every year. Plague and typhus are enzootic in wild rodent populations in some areas, especially in the warmer climates. Thus there are sporadic outbreaks of human cases even in the United States. Transmission to commensal rat populations from wild rodents is always a possibility, making continuous rat control important.

Among other diseases transmitted by rats to man, leptospirosis, rat-bite fever, and food poisoning from *Salmonella* organisms are especially important.

Wild rodents of many kinds may be involved in the transmission of disease. Prominent among these are ground squirrels and wood rats which transmit spotted fever through ticks that come into contact with man. Many rodents are susceptible to tularemia infection, but rabbits and hares are more often handled and thus are more frequently involved in tularemia transmission to man.

Control

Rat-borne diseases are controlled most effectively by measures that reduce rat infestations. These are primarily sanitation (removal of rat food), ratproofing (removal of rat shelter), and destruction of the rats.

In controlling plague and typhus, it is advisable to dust rat harborages and burrows with an appropriate insecticide before killing the rats, in an effort to destroy rat fleas, the disease vectors that otherwise might move directly to man when their normal hosts are killed.

Control of rodent-borne diseases through reduction of wild rodent populations is not a practical solution except in very special local situations. When disease is known to be present, direct contact with the animals or their parasites (mites, fleas, ticks) should be avoided. Area dusting or spraying with insecticides in such situations as military encampments helps destroy the parasite vectors. Repellents applied to skin and clothing provide individual protection.

Assessment

Rodent-borne diseases transmissible to man are present in many parts of the world. In the more highly-developed countries, sanitation, rat and vector control, and suitable precautions have greatly reduced the incidence of these diseases in man and domestic animals. In underdeveloped countries, these diseases still result in much economic loss, human misery, and death.

PROBLEMS WITH SMALL CARNIVORES

Rabies is the outstanding disease associated with carnivores that is transmissible to man and domestic animals. This dreaded disease is enzootic in many regions of the world, and epizootics occasionally erupt among dogs, wolves, coyotes, foxes, shunks, and other carnivores. During such outbreaks in wild species, the disease may be transmitted to man or livestock when the disturbed behavior patterns of sick animals bring them into direct contact.

Red or gray foxes are involved most often in rabies outbreaks in the United States, although in some areas coyotes or shunks play the major role.

Major economic losses occur during epizootics when cattle are bitten, contract the disease, and must be destroyed. The annual costs due to rabies in the United States may total an estimated $5 million or more in livestock losses and the treatment of more than 30,000 persons exposed to the disease (Held *et al.*, 1967).

Control

Diseases such as rabies in wild carnivore populations are probably self-limiting in that all infected animals die. In local outbreaks attempts often are made to hasten population reduction by trapping, poisoning, or other means, but this seldom succeeds completely in halting the outbreaks.

The most practical rabies control method is annual mass vaccination of dogs, strictly enforced licensing of all dogs, and humane disposal of all unlicensed or stray dogs. Cattle also may be immunized against rabies, but it is sometimes less expensive to insure the herd against loss.

Assessment

The possibility of reducing excessive populations of wild carnivores before rabies infection spreads to them is an attractive concept, since the disease appears to be somewhat dependent on the density of the population. Research is in progress on population control through suppression of reproduction by chemical means.

The possibility that highly mobile bats serve as reservoirs of rabies infection is disturbing and may explain the recurrence of rabies outbreaks in carnivore populations. This bat-carnivore relationship is not yet established clearly.

PROBLEMS WITH BATS

Most bats of the United States, because of their voracious appetites for insects, are normally beneficial and should be protected from wholesale destruction. Some species are also valued for their guano deposits, at least where large roosting areas are exploitable.

On the other hand, some bats and their arthropod parasites are known (and others suspected) to transmit various viral, parasitic, and fungal diseases to man, livestock, and wildlife. Among these are rabies, histoplasmosis, Chagas' disease, relapsing fevers, dermatomycoses, and Japanese B encephalitis. The bat's role in rabies transmission is the most serious of these health hazards because bats can carry and transmit the virus with or without developing the clinical symptoms of rabies. Thus they serve as reservoirs for the disease, transmitting it directly to other animals or man by biting (if the virulent form of the disease is contracted) or indirectly, when rabid bats are eaten by predators that become infected with the disease and bite other animals or man.

Although opportunity for transmission of pathogens is afforded by close sharing of many of the same habitats between bats, men, and animals, the significance of insectivorous bats—the only group of immediate concern in northern North America—in disease problems remains to be clarified. At present, their greatest offense seems to be daytime roosting in occupied dwellings or other buildings. The repugnant odors of their excretions and the distracting noises they make inside walls or while crawling in and out of established roosts often warrant their control.

Because rabies is infectious to mammals in general, most bat species found in the United States may be assumed to be potential rabies carriers. Bats known to roost in buildings include members of the genera *Myotis, Eptesicus, Pipestrellus, Tadarida, Antrozous, Lasionycteris,* and others.

There has been an apparent increase in the incidence of bat-transmitted rabies in the United States in recent years. In 1953, for example, there were 8 cases reported in man; in 1962, there were 157 (U.S. Public Health Service, 1962). This increase reflects, in part, human population increases, increased opportunity for contact with bats, and increased emphasis on detection of diseases. In most areas, the risk of rabies infection from bats is slight, but in some, the opportunity is high for contact between rabid bats and human beings.

There is recent evidence (Constantine, 1962) that bat rabies is transmissible by an airborne medium, such as a suspension of particles in air, under certain conditions occurring in bat caves. Whether there is a risk of aerosol transmission in the vicinity of other types of roosts is not known.

Other bat-borne diseases, including histoplasmosis and coccidioidomycosis, have been identified mainly in persons such as cave explorers and guano miners who have spent considerable time inside bat caves.

Bats have also been a problem in air strikes with fast jet aircraft (e.g., T-38 jet trainers).

Control

In view of the beneficial aspects of bats, major control campaigns in the United States are seldom excusable. Only during epidemics of rabies or other bat-borne diseases is there justification for widespread punitive action. Control is necessary in localized nuisance situations, such as bat roosts in homes.

Little information is available regarding economic losses due to bat damage. Control expense must, of course, be based on the specific situation and economic loss involved. The relative costs and advisability of bat-proofing or using repellents or lethal agents depend on factors too variable to consider in general. However, where a hazard to human health is involved, the cost of control is seldom an important factor.

Bat roosts in occupied buildings can be eliminated by killing the bats, repelling them with chemicals, or bat-proofing the building. Killing provides only temporary relief, because the attractiveness of the habitat remains, as do the roost odors, inviting other bats to reestablish the colony. If the roost is inaccessible (between walls, for instance), the bat carcasses cannot be removed, and decomposition smell only adds to the unpleasant odors already associated with the roost. Bats dead or dying of intoxication from pesticides may fall to the ground near the building, increasing the risk of human exposure to bat-borne diseases.

Since bats eat only insects, treated baits cannot be used as an eradication method. Toxicant formulations liberally dusted on and near the roost areas and entrances to house roosts are sometimes effective because the bats may ingest lethal amounts of the material while grooming. Toxic compounds used successfully for this purpose include DDT and dieldrin. Fumigation with hydrocyanic acid gas has also been used effectively for bat destruction, but is dangerous to human beings, requiring that the building be vacated. This method is subject to legal restrictions, and should be employed only by an experienced operator.

Repellents may be useful in keeping bats out of a building, but like toxicants, their effect is temporary, and treatments must be repeated every few weeks. Benzene hexachloride (BHC), chlordane, paradichlorobenzene, naphthalene, mustard oil, and bird and rodent repellents have been used. A more permanent repellent in some cases is fiber glass insulation, which can be blown into spaces occupied by bats.

The most satisfactory way to eliminate house roosts is to bat-proof the building. Where only a few avenues of entry are available to bats, the process is quite simple, but in warm climates the style of architecture may prohibit effective bat-proofing. For any type of construction, the work must be thorough because the smaller bats can enter openings no more than ¼ inch wide. Cracks and crevices larger than this should be eliminated from walls, roofs, and floors. Bat-proofing should be completed in two stages: sealing all but one or two of the principal openings at first, then waiting a few days until the bats learn to use the remaining crevices before closing them. If the work is done after the bats swarm out in the evening, all bats should be trapped outside when the last opening is sealed. The cracks may be closed with strips of metal or wood, or they may be plugged with rags, oakum, or other soft material.

Assessment

There is no known practical way of assessing the public health aspect of bat-borne diseases. It is a sporadic and widespread problem and must be dealt with locally as the need arises. Bat roosting in houses is also a sporadic situation, but more common in eastern and southern United States. Actual damage figures are not available at this time.

During outbreaks of rabies, federal, state, and local health authorities assume responsibility for directing the control program, which often involves other wild mammals as well as bats. Federal and state control specialists and commercial pest control operators often perform or supervise the actual control work.

There is a great need for safe, efficient means of bat damage control. Toxicants currently used leave much to be desired, and available repellents are mostly unsatisfactory. Good methods of applying toxicants are lacking, and secondary hazards have not been delineated. Investigation of bat damage control could well be conducted by both government and university researchers.

Beyond these problems, it should be emphasized, especially to the public, that the insectivorous bats are generally very desirable from the human viewpoint, and that control measures should be undertaken only in local situations involving serious damage or health hazards.

SOME SPECIAL PROBLEMS

The introduction of non-native animals into a country or region may create special problems. The animal in a new environment may change its habits and destroy valuable crops, it may displace or attack native animals, and it may carry or introduce diseases new to the region.

There are many examples of this type of problem (de Vos *et al.*, 1956), such as the introduction of the European rabbit to Australia and New Zealand, the mongoose to the West Indies and Hawaii, the muskrat to Europe and Asia, the American gray squirrel to England, and the European hare to Ontario and New York, to name only a few.

NUTRIA

A special case may be cited concerning the introduction of the South American nutria or coypu into North America (U.S. Fish and Wildlife Service, 1957). These large rodents were imported to fur farms in the United States and Canada, from which they escaped or were released when the projects failed. Nutria have survived in the wild in many parts of the United States, and in some areas have increased greatly in numbers to reach the status of a pest. Because they feed on much the same vegetation as muskrats, they have displaced this valued furbearer over large areas of marshland along the Gulf Coast and the lower Mississippi River. However, nutria are harvested for their fur and meat (for ranch mink food) and this partly compensates for reduced muskrat numbers (O'Neil, 1965).

A wide variety of farm crops that are grown close to waterways occupied by nutria may be damaged. The animals also destroy much habitat for migratory waterfowl in some states by eliminating all vegetation over large areas of marsh.

Control measures against nutria are used in areas such as wildlife refuges (U.S. Fish and Wildlife Service, 1967). Zinc phosphide baits were used successfully on an experimental basis. Anticoagulant baits have also been used, but secondary poisoning has occurred in mink and dogs fed nutria meat from such animals (Evans and Ward, 1967).

ARMADILLOS

The nine-banded armadillo is another example of a controversial introduction. While extending its range naturally into some states, it has been introduced into others, such as Florida. Its burrows may undermine buildings, damage gardens, and penetrate dikes and levees, causing leakage and washouts. Ground-nesting birds such as quail may be affected by excessive numbers of armadillos. The burrows are of some value as refuges for other animals, and the horny shell is utilized for novelties to be sold to tourists. The usual food of armadillos is insects and other invertebrates (Fitch *et al.*, 1952).

One useful method of control is night hunting with lights and clubs.

APPENDIX: SCIENTIFIC NAMES OF NORTH AMERICAN SMALL MAMMALS

Common Name	Scientific Name	Common Name	Scientific Name
Armadillo	*Dasypus novemcinctus*	Norway rat	*Rattus norvegicus*
Beaver	*Castor canadensis*	Nutria	*Myocaster coypus*
Big brown bat	*Eptesicus fuscus*	Pallid bat	*Antrozous pallidus*
Black rat	*Rattus rattus*	Pine mice	*Microtus* (*Pitmys*) spp.
Chipmunk	*Tamias* and *Eutamias* spp.	Pipistrelle bat	*Pipistrellus* spp.
Cotton rats	*Sigmodon* spp.	Pocket gophers	*Geomys, Thomomys,* and *Cratogeomys* spp.
Coyotes	*Canis latrans*		
Dogs	*Canis familiaris*	Pocket mice	*Perognathus* spp.
Free-tailed bat	*Tadarida* spp.	Porcupines	*Erethizon dorsatum*
Gray fox	*Urocyon cinereoargenteus*	Prairie dog	*Cynomys* spp.
Ground squirrel	*Citellus* spp.	Rabbit	*Sylvilagus* spp.
Hares and jack rabbits	*Lepus* spp.	Raccoon	*Procyon lotor*
House mice	*Mus musculus*	Red fox	*Vulpes fulva*
Kangaroo rat	*Dipodomys* spp.	Rice rat	*Oryzomys* spp.
Little brown bat	*Myotis* spp.	Shrews	*Sorex* spp.
Marmots (woodchucks)	*Marmota* spp.	Silver-haired bat	*Lasionycteris noctivagans*
Meadow mice	*Microtus* spp.	Shunks	*Mephitis* and *Spilogale* spp.
Moles	*Scalopus, Parascalops, Condylura, Scapanus, et al.*	Tree squirrels	*Sciurus* and *Tamiasciurus* spp.
		White-footed mice	*Peromyscus* spp.
Mountain beaver	*Aplodontia rufa*	Wolves	*Canis lupus*
Muskrat	*Ondatra zibethicus*	Wood rats	*Neotoma* spp.

Source: Hall and Kelson (1959).

REFERENCES

Anonymous. 1968. Rats are stymied by new plywoods. New York Times, Sept. 15, 1968, p. 44.

Bailey, V. 1924. Breeding, feeding and other life habits of meadow mice (*Microtus*). J. Agric. Res. 27:532–535.

Besser, J. F., and J. F. Welch. 1959. Chemical repellents for the control of mammal damage to plants. N. Amer. Wildl. Conf. Trans. 24:166–173.

Constantine, D. G. 1962. Rabies transmission by nonbite route. Public Health Rep. 77(4):287–289.

Crabtree, D. G., W. H. Robison, and V. A. Perry. 1964. Compound S-6999 (McN-1025), new concept in rodent control. Pest Control May 1964:26–32.

deVos, A., R. H. Manville, and R. G. Van Gelder. 1956. Introduced mammals. Zoologica 41(4):163–194.

Eadie, W. R. 1959. Toxaphene for control of meadow mice in orchards. Cornell Univ. Agric. Exp. Sta., Farm Res. 25(1). 15 pp.

Evans, J., and A. L. Ward. 1967. Secondary poisoning associated with anticoagulant-killed nutria. J. Amer. Vet. Med. Ass. 151:856–861.

Fitch, H. S. 1948. Ecology of the California ground squirrel on grazing lands. Amer. Midl. Nat. 39:513–596.

Fitch, H. S., P. Goodrum, and C. Newman. 1952. The armadillo in the southeastern United States. J. Mamm. 33:21–37.

Hall, E. R., and K. R. Kelson. 1959. The mammals of North America. The Ronald Press Co., New York. 2 vol.

Held, J. E., E. S. Tierkel, and J. H. Steele. 1967. Rabies in man and animals in the United States, 1946–65. Public Health Rep. 82:1009–1024.

Horsfall, F., Jr. 1953. Mouse control in Virginia orchards. Virginia Agric. Exp. Sta. Bull. 465:1–26.

Kverno, N. B. 1964. Forest animal damage control. Proc. Vertebrate Pest Control Conf. Anaheim, California, pp. 222–226.

Miller, M. A. 1953. Experimental studies on poisoning pocket gophers. Hilgardia 22:131–166.

O'Neil, T. 1965. Fur future. Louisiana Conserv. 17(3+4):14–17.

Rudd, R. L. 1964. Pesticides and the living landscape. Univ. Wisconsin Press. 320 pp.

Shaw, W. T. 1920. The cost of a squirrel and squirrel control. State Coll. of Washington, Agric. Exp. Sta. Bull. 118:1–19.

U.S. Fish and Wildlife Service. 1957. Nutrias in the United States. U. S. Fish and Wildlife Service, Wildlife Leaflet 389. 11 pp.

U.S. Fish and Wildlife Service. 1967. Wildlife research; problems, programs, progress, 1966. U.S. Fish and Wildlife Service, Resource Publ. 43. 117 pp.

U.S. Public Health Service. 1962. Annual Rabies Surveillance Report for 1961. U.S. Public Health Service, Communicable Disease Center, Rabies Control Unit. 13 pp.

Vertrees, J. D., and D. A. Spencer. 1959. The Oregon meadow mouse eruption of 1957–1958. Oregon State Coll. Corp. Est. Serv.: 1–88.

Ward, A. L., and R. M. Hansen. 1960. The burrow builder and its use for control of pocket gophers. Special Scientific Rept., Wildlife No. 47. U.S. Fish and Wildlife Service. 7 pp.

CHAPTER 6

Predatory Mammals That Become Pests

Predation is the natural killing of one animal by another. More often than not the victim is eaten in order for the predator to survive. For the welfare of the predator, the prey, and the environment they coinhabit, such deaths are vital to the food-chain cycle.

Predators in the natural environment must remain considerably fewer in numbers or lower in biomass than the prey, and they must have lower recruitment rates. When this relationship fails, predator populations decrease until a balance with the animal food supply is reached. Some of the components of complexity in dealing with predation as a biological phenomenon are the following: within-species interaction, "buffer" alternatives (Leopold, 1933), compensatory mechanisms (Errington, 1967), predator and prey densities, cover conditions, human exploitation, prey food preferences, obligate or semi-obligate predation, weather and season, physical condition of predator and prey, disease, and human error in observation or assessment.

The environment (soil, plants, weather, terrain) not only dictates the kinds and numbers of animals to be supported, but also the kind and amount of predator–prey interaction that will adjust animal populations to the capacity of environment to support them.

Apart from environment, but influenced by it, are the behavioral traits of the predators and prey. Territoriality, migratory urge, aggressiveness, gregariousness, inquisitiveness, etc., are a part of the dynamics of living and dying among animals. Basic to any conceptual approach to predation is the fact that all predators themselves become prey at times, and thus predator–prey adjustment is not always flesh-eater reducing plant-eater. Indeed, the balance with food supply is reached more rapidly when predator reduces predator, and the surviving plant-eaters are thus buffered by other flesh-eaters.

In this complex of interactions there may be times when it is necessary to control predators. To do this adequately, without danger to other forms of life and in a manner compatible with the welfare of man, is no easy task. Skill, knowledge, and judgment are required.

If animals living in a given place are arranged in a pyramid of numbers or of biomass, the flesh-eaters are at the top and the more abundant plant-eaters at the bottom. In the broader ecological areas, there may be smaller pyramids within the larger pyramid, but the relationship of plant-energy converters to animal-energy converters is always the same. Wherever a naturally or unnaturally induced imbalance occurs in the numbers of predator or prey species, the stability of the entire pyramid is affected. This imbalance causes one or the other to become a pest.

In pristine times there were no pest situations. Natural imbalance, when it occurred, righted itself in time. Biological control was the only force available to adjust the compatibility of flesh-eaters, plant-eaters, and their environments. These adjustments to imbalance constantly test the plants and animals of an ecosystem, forcing them through physiology, structure, and behavior to adapt to change brought about by biological imbalance. Permanent biological balance is rare, if it ever exists, in the wild. The sum total of these adjustments comprises the tangible evidence of evolution. Those organisms large and small that failed to adjust when rapid, large-scale imbalance was thrust upon them are now only records in the annals of paleontology.

Because fossil evidence shows that man and predators lived side by side in the distant Pleistocene, we must assume an interaction involving these two life forms. It is likely that they preyed on each other; cave paintings tell of this conflict. Recorded history is more flamboyant in overstating the case against the flesh-eaters as adversely affecting man's welfare. Under a welfare banner, man waged an effective war of extinction on many carnivores. Only after the predator is gone are the process and the result assessed with some semblance of rationality. The assessment frequently uncovers waste, greed, and guilt on the part of the eradicators.

America's Indian population lost horses and domesticated animals (e.g., turkeys) to predators and, after the arrival of white men, risked the hazards of rabies from dogs, foxes, wolves, and other mammalian predators. The colonists were not exempt from wild animal predation when they first arrived with their livestock. The major conflict was to come later when early stockmen encroached on virtually all range occupied by major predators. Disrupted food chains and restricted range caused the flesh-eaters to turn to the most available source of food, livestock.

This basic confrontation was the genesis of our present-day dilemma involving the reconciliation of survival for wild predatory animals and of protection for domestic protein and animal-fiber production.

THE MAJOR PREDATOR PROBLEM

Animals that prey on livestock—on cattle, horses, sheep, swine, and poultry—constitute the major predator problem. Because their predations result in great economic loss, they attract the most attention and even arouse political furor.

Livestock were first introduced into the American colonies in 1610 and depredation, particularly on sheep, began at once. By 1650 cattle were to be found in all American colonies, and J. A. Allen (1876) records, "Most of the carnivorous species existed in such numbers at the time of the first settlement of the country by Europeans that their presence was a great check upon the raising of stock, and even a source of danger to human life." Although other predators were undoubtedly active, wolves were most numerous and destructive. One assessment of this history of depredation is given by S. P. Young (1946).

The natural and immediate reaction was to destroy the predator: as an incentive the first bounty was offered by the Plymouth colony in 1630, and steel traps were introduced about 1645 to combat the common menace. The larger predators such as wolf and puma disappeared with the inroads of civilization, but the smaller carnivores remained to plague the landowner. The problem of depredation by the large predators continued westward as man encroached on the natural ranges of these animals.

In the early West, horses and mules fell victim to these predators, and the losses were increased with the coming of meat animals. Railroads spelled the demise of the bison, and wolves turned to livestock as a substitute. Coyotes, wolves, and puma made sheep raising on the frontier risky. It is not surprising, therefore, that livestock owners cooperated in killing the offending animals. Initially the costs of such a system fell to those who benefited from the "control," but through political pressures this obligation was gradually foisted onto the county, state, and finally the federal government. In the meantime, private associations and local governments paid well enough so that some individuals became professional bounty hunters. In 1915, when the first federal monies became available for predator control, there was a cadre of trappers and hunters available for work on public lands on a salary basis. For the next 15 years federal appropriations increased until 1931, when a bill was passed authorizing an amount not to exceed $1 million annually for cooperative predator control on public and private lands. Additional monies came from private groups and local governments.

Today the wolf has been replaced by the coyote and the puma by the bobcat as prime targets for mammal predator control. The methodology as well as the need for predator control are criticized by some as excessive. One of the most cogent arguments (apart from the biological) against governmental participation in predator control is the economic consideration. In some cases the cost to government for protecting cattle and sheep exceeds the value of the livestock lost (Table 1), although it might be argued that if there were no control the predators would increase in numbers, and the losses would likewise increase. In the case of sheep and coyote, this is not a direct cause-and-effect relationship. Instead, both predator and prey seem to fluctuate in synchrony. The reason for the pattern, as alleged by A. G. Etter (U.S. House of Representatives, 1966), is that "... it would appear that the take of coyotes has been more of an indication of the numbers of sheep on the range rather than of coyotes, simply because the funds to trap and poison are dependent on numbers of assessable sheep." This situation is further clouded by the fact that in addition to tariffs on meat and wool, the government is asked to support a sheep-protection program whose merits are frequently and severely challenged. In addition, about two million sheep now graze on national forests alone (U.S. Department of Agriculture, 1967).

It is illogical for the government to support a private commercial endeavor where the cost of protection exceeds the value of that which is protected and where the protection program conflicts with other publicly owned natural resources.

When all the polite considerations are stripped from the controversy of livestock depredation, there remains the confrontation of the coyote as the pred-

TABLE 1   Sheep Grazing, Predator Losses, and Costs of Predator Control in Four Western Regions of the U.S. Forest Service for 1962

USFS Region	Sheep Grazed[a]	Sheep Lost[b]	Percentage of Sheep Lost	Value of Sheep Lost[c]	Cost of Control[d]
I	153,788	1,435	0.9	$ 24,784	$ 20,044
IV	1,143,219	12,630	1.0	218,120	142,902
V	64,743	223	0.3	3,501	90,195
VI	116,223	1,116	0.9	19,273	37,908
Total				$265,678	$291,049

[a]For average period of 3 months in summer.
[b]All losses charged to predators.
[c]From Statistical Bulletin Nos. 333 and 230, USDA Agricultural Marketing Service.
[d]Cost of control work on forest lands.

Source: The Leopold Report; Leopold, 1964.

ator and the sheep as the prey. All socioeconomic conflicts of other species scale down from this end of the predator-prey spectrum.

To assess this situation, the following facts and positions must be understood and placed in proper perspective:

• The sheep industry is declining on a national scale, e.g., 4 percent for the year ending January 1, 1967 (Colorado Wool Growers Association, 1967). According to Young and Jackson (1951) ". . . sheep in the United States were in liquidation beginning in early 1942, shortly after our entry into World War II. This reduction was estimated at about 7 percent a year to 1946, when it was ascertained to be nearly 9½ percent."

• The cost of predator control for sheep exceeds the on-the-hoof value of the livestock killed by predators (Leopold, 1964).

• The federal government subsidizes the livestock industry by allowing sheep and goats to graze on public lands. As of 1968, the average charge per animal unit month on Bureau of Land Management lands was 33 cents, and on National Forests 55 cents, compared with "a fair price" of $1.31 and $1.50, respectively, as determined by an official Bureau of the Budget report released late in 1968. (In January 1969, the retiring Secretaries of Agriculture and of the Interior directed that the grazing fees on these public lands be increased over a 10-year period until they reached the "fair price.") Other federal subsidies to the livestock industry include tariffs and import restrictions, incentive payments to wool producers, and predator control.

• Sheep raising is of two kinds: as an animal crop in diversified farming (usually in the eastern half of the country); and as free-ranging sheep in single-crop ranching (in western United States). It is the rancher who requests predator control.

• Predator loss (including loss from eagles) cannot be accurately determined in most cases. Claims of stockmen that have been verified are usually few.

• Not all coyotes (or eagles) take sheep. Coyotes scavenge on carcasses of sheep dead or dying from natural causes.

• The use of herders and of lambing sheds is apparently too expensive for most sheep raisers, but losses could be cut considerably by this change in husbandry. Herding, however, tends to concentrate grazing pressure on restricted areas causing destruction of forage plants and producing trails that cause erosion.

• Methods of predator control are at times excessive and nonspecific and are leveled against species that are part of a national natural resource complex.

• Lethal poisons are available to the nonprofessional to use as he may in predator control.

• The National Wool Growers' Association is the most active agency pro-

moting the use of federal funds for predator killing; the Defenders of Wildlife
and the National Audubon Society are the mainstays in protecting target and
nontarget species from attempts at predator control.

• In 1965, it cost the government and cooperative agencies $5,614,806 for
the predator control program alone. In the same year, 90,236 coyotes were
killed along with other predators taken accidentally or purposefully in the
process. How many of the coyotes taken were not sheep predators is unknown.

• Sheep cared for by herders and sheep dogs suffer little mortality compared with unattended, free-ranging flocks. The lambing period is a crucial
time when protection by herders is necessary for maximum production.

From all this there too often develops controversy between those seeking
financial aid to kill flesh-eaters in the name of local economy, and those seeking protection for the predators under the banner of national conservation.

The most elusive pieces of information necessary to sound judgments are
those concerning costs of control efforts and of animal losses. If the predator
control program is of sufficient import to the public, government, and the
private citizen then it is imperative that these data be accurate and readily
available. Predator control agencies must arrange their bookkeeping so that
costs of predator control and the necessary itemizations to make the record
understandable are not lost in a welter of other financial transactions. At the
same time those agencies recording losses of domestic animals to predators
must make certain that the statistics are accurate. There is always the tempta-
tion to attribute larger losses to predation than are warranted (S. A. Cain in
U.S. House of Representatives, 1966, p. 34).

Predation losses to sheepmen and livestock owners are very real. If the
methodology and results of current predator control are unacceptable socially
and biologically, it becomes incumbent on government to aid in developing
adequate substitutes, and this special interest group should cooperate in the
research effort. However, there is little evidence of joint effort in the government-sponsored predator control program.

In an effort to clarify the situation, Secretary of the Interior Stuart Udall
asked his advisory board on wildlife management in 1963 to address itself to
"predator and rodent control in the United States." The Leopold Report
(Leopold, 1964) was the result.

THE LEOPOLD REPORT

The key federal agency in the predator control program is the Division of
Wildlife Services (formerly the Branch of Predator and Rodent Control) in the
Fish and Wildlife Service. It contributes about 40 percent of the approximately

$6 million expended annually. The remainder is contributed by other federal agencies, counties, states, and livestock associations.

Two basic tenets were adopted by the inquiry board as a basis for recommendations:

1. All native animals are resources of inherent interest and value to the people of the United States. Basic governmental policy therefore should be one of husbandry of all forms of wildlife.

2. At the same time, local population control is an essential part of a management policy, where a species is causing significant damage to other resources or crops or where it endangers human health or safety. Control should be limited strictly to the troublesome species, preferably to the troublesome individuals and, in any event, to the localities where substantial damage or danger exists.

The group was also unanimous in the opinion that "[predator and rodent] control as actually practiced today is considerably in excess of the amount that can be justified in terms of the total public interest." This is one of the most significant statements in the entire report, because it focuses attention on the reasons for the excess, the methodology resulting in excess, the bio-social meaning of excessive control, and the responsibility for guiding a program of excessive control.

The subsequent assessment of programs that were "complicated almost beyond belief" condemned parts and approved parts of the government's activities in control. Subjects also covered were the damage caused by predators; the hazards of using the very lethal poison 1080 (sodium fluoroacetate), particularly with respect to secondary poisoning; rabies outbreaks; rodent and bird depredation and control, particularly the role of the golden eagle and sheep; and research on control methods. Deliberations on these and associated aspects of the problem resulted in the following recommendations:

1. An advisory board on predator and rodent control should be appointed to the Secretary of the Interior.

2. The Branch of Predator and Rodent Control should completely reassess its function and purpose in the light of changing public attitudes toward wildlife.

3. With respect to predator and rodent control operations, it was suggested that

    (a) Explicit criteria are needed to guide control decisions;

    (b) There should be continued cooperative programs with other agencies;

    (c) Proof of control need (documentation) is necessary;

    (d) Extension trapper programs should replace bounty schemes; and

(e) Flying squads of control agents should be established to cope with rabies outbreaks (in eastern United States).

4. There should be a greatly amplified research program for specific controls and nonlethal control devices.

5. Legal measures are needed to regulate the use of poisons in control operations and to prevent shipment of such poisons to foreign countries lacking adequate regulation of them.

The report concluded with the notation that control programs should be guided by demonstrated need and scientific management, although those seeking federal assistance in the field may well find it difficult to demonstrate need by objective criteria. There are as yet no published guidelines to eliminate or minimize subjective judgment and its attendant biases.

However, the impact of general policy documents such as the Leopold Report is unlikely to be such as to bring about rapid or dramatic changes in the face of public tradition and established operational procedures.

After almost five years of the "new look" in federal predator and rodent control, there is both criticism and progress. An advisory board can set the stage for efficient operation, but unless there is built into the program a system of feedback to test the recommendations and make alterations, private watchdog organizations serve as the only evaluators.

A far better use of an advisory board, particularly to an agency responsible for animal control, would be to have the board evaluate and adjust its recommendations once field results are obtained. In many cases, advisory boards move from one problem to another without having an opportunity to check the results of recommended action on a particular problem.

If there is no vehicle for feedback, the public through militant groups will react forcibly, sometimes with bias and virtually always with emotion. This is not to say that such reaction is without merit—quite the contrary—but the public's reaction may be direct or devious, and to all levels of administration, often the very top.

A case in point is the use of poison bait (1080) stations in western United States. The Leopold Report recommended one such station per township as reasonable and adequate for coyote control. Etter (1968) describes a situation in northwestern Colorado that appears to flaunt the recommendation and demands an explanation of demonstrated need. In 191 townships recording bait stations 83, or 43 percent, had one such station; in the other 57 percent, the number of stations ranged from 2 to 15 and averaged 3.4 per township. This situation has since been corrected.

This is only one case of failure to implement a recommendation in the field. On the whole the Division of Wildlife Services has been successful in implementing the Leopold Report recommendations, such as the upgrading of personnel, the redirection of programs, effecting a closer liaison between Washington and field stations, and improving public relations.

When advisory board recommendations result in major field problems, these problems should be returned to the board itself for reassessment and possible alternative action. In this manner, the chief administrator can obtain the maximum benefit from the board of specialists best able to appraise the entire problem. Boards that recommend and then are dissolved often are unaware of the adequacy of their guidance. A follow-through board will have a broader charge, but its chances of producing a successful action program are increased.

OTHER PREDATOR PROBLEMS

Predation by the larger carnivorous mammals does not cover as wide a geographic area nor does it create as great a financial loss as does coyote predation. Wolves and puma still kill livestock where they are in sufficient abundance and where their ranges overlap with those of livestock. Bear depredations also create problems in the East and West. Livestock raising is declining in the remote areas of the East and Midwest, so the opportunity for conflict has been reduced. Skunk and raccoon predation on poultry and gamebirds remains, but varies greatly in location and intensity; the same is true of otter and mink predation on fishery resources. Each problem situation is the result of variables specific to the circumstances in the locality. One can only generalize by recommending safe methods for control of the individuals causing damage to the resource. The obvious danger lies in instituting a control program on the species as a reaction to depredation.

Exaggeration of the predation by red and gray foxes on rabbits, poultry, and pheasants in folklore, fairy tales, and sport magazines has conditioned many people to uncritical assessment of foxes as wild animals. In frontier days when man diversified the landscape with farms, small farm animals became vulnerable to foxes. Today, intensive agriculture and predator-proof housing for poultry have virtually eliminated barnyard raids by foxes. Food-habit studies of the fox indicate that domestic stock has a minor role in its diet. In cases where evidence of predator is lacking, a fox is usually considered to be the culprit. If the truth were known, raccoons would doubtless share heavily in the blame.

Foxes do take some poultry, young farm mammals, and game animals, but their primary diet is field rodents. This applies to foxes in virtually their entire geographic range. There are specific cases where foxes should and must be controlled. However, to alleviate a depredation problem by a general assault on all animals of the species is wasteful, uneconomic, and does not solve specific problems. Such problems can only be solved by controlling or eliminating the offending individuals. Table 2 identifies some of the predator–prey conflicts involving control.

PEST CONTROL PROGRAMS

FEDERAL

The Leopold Robert (Leopold, 1964) fully describes the current federal predator program and modifications resulting from the report are indicated by Gottschalk (1967). Under the government program that has evolved since 1915 the Division of Wildlife Services, Bureau of Sport Fisheries and Wildlife, U.S. Department of Interior, furnishes less than half (currently about 40 percent) of the total cost, with other governmental agencies, state or local governmental units, or private livestock organizations furnishing the greater share. The Division of Wildlife Services provides the professional staff, including supervisory personnel, and the Division of Wildlife Research conducts research into safer and more effective control methods. The control work is conducted largely by full-time professional operators. Possible alternatives to full-time professional operation will be considered below.

THE BOUNTY SYSTEM

The easiest way to wage war on a predator is to put a price on its head—this is called the "bounty system." The objective is to protect farmers from losses of livestock and to increase wildlife by eliminating what is thought to be man's chief competitor in the use of that resource. There is ample evidence that this system does not accomplish its objective. Young and Jackson (1951) wrote an excellent appraisal of the bounty system; Arnold (1956) evaluated the system as it operated in Michigan.

The bounty system is still used today, although it has been shown to be

Ineffective. "For instance, in 1935 when the present Michigan bounty on wolves and coyotes began, 27 wolves and 3,026 coyotes were bountied. In 1954, after the expenditure of $10,215 for wolf bounties and $912,510 for coyote bounties, 23 wolves and 3,743 coyotes were bountied. In 1948(?) some 20,968 foxes were bountied and in 1954 after $801,140 had been spent, trappers presented 26,964 foxes for bounty. One can hardly argue that these animals are being eliminated or even reduced in number" (Arnold, 1956).

TABLE 2 Predation Damage Resulting in Requests for Control

Species	Depredation	Time	Degree
Coyote	Sheep, lambs, goats, and kids	All year	Moderate to severe
	Calves	All year	Slight to moderate
	Poultry and big game (yg.)	All year	Slight
	Melons	Sp	Slight
Wolf (red)	Calves	Sp	Slight
Wolf (gray)	Livestock	All year	Slight
	Big game	All year	Moderate
Fox (gray)	Lambs, kids, and pigs	FWSp	Moderate
	Poultry	All year	Slight
Wolverine	Trapped fur animals, trappers' caches	FW	Slight
Bear (grizzly)	Sheep and livestock	SF	Moderate (drought)
Bear (black)	Sheep	SF	Moderate (drought)
	Pine seedlings	SpSF	Slight
	Apiaries	SpSF	Slight to moderate

Raccoon	Lambs, kids, poultry, truck crops	All year	Moderate
	Ground and tree nesting birds	SpSF	Slight to moderate
	Stored grains	FW	Slight to moderate
Weasel	Poultry	All year	Slight to moderate
Mink	Poultry	All year	Slight to moderate
Skunk	Poultry	SpSF	Slight to moderate
Badger	Poultry and lambs	SpSF	Slight
River otter	Fish (hatcheries)	All year	Slight
Mountain lion	Sheep and lambs	All year	Moderate
	Goats, kids, horses, colts, big game	All year	Slight
	Livestock	Unusual	Slight
Bobcat	Lambs	SpS	Slight to moderate
	Kids, poultry	All year	Slight to moderate
	Big game (yg.)	SpS	Slight to moderate
Jaguar	Livestock	All year	Slight
Ocelot	Lambs and kids	SpS	Rare

Source: This evaluation, based on long experience in predator control work, was made by Milton Caroline, Texas State Supervisor, Division of Wildlife Services, U.S. Fish and Wildlife Service.

Fraudulent. In Michigan, Douglas and Stebler (1946) wrote: "During these 'bountiful' years there developed a flourishing business of importing pelts of 'wolves' (mostly coyotes, which were not distinguished from timber wolves in the bounty payments) from western and southwestern states. It wasn't a bad investment, buying Dakota coyote pelts for around $3 and selling them to the state of Michigan for $35. One pair of operators submitted 248 pelts to western Northern Peninsula counties for payment, and collected $8,680 for nine months' trafficking.

"At the same time that large-scale skulduggery was going on in the big-money wolf and coyote bracket, small-time chiseling was rampant with the lesser species. . . ."

Young and Jackson (1951) added: "Human nature being what it is, as soon as an objective (killing of predators) is made a source of income (by payment of bounty) the prevailing idea becomes one of propagating rather than eradicating."

Wasteful. From the Leopold Report (1964): "In the opinion of this Board, far more animals are being killed than would be required for effective protection of livestock, agricultural crops, wildland resources, and human health. This unnecessary destruction is further augmented by state, county and individual endeavor." Many wasteful aspects are not easily documented, particularly nonbountied animals killed in the trapping process. In some cases, sportsmen's dollars are used to support bounties more than would be economically justified. In Michigan, for example, Shapton (1951) wrote: "Do you know— That Michigan hunters are actually paying $19.38 for each fox removed under the bounty system rather than $5 as the law specifies? Though 75,509 foxes have been turned in for bounty, it has been conservatively estimated that at least 55,000 foxes would have been taken anyway through normal hunting and trapping. The state bounty, therefore, actually removed only 20,509 foxes at a cost of $397,545, or about $19.38 for each fox over and above what would have been taken had not the bounty been in effect."

Costly. Bounties are still paid by a number of state and local governments. In addition to state and local bounties, the federal government in 1965 paid $6.9 million for predator, rodent, and bird control activities (U.S. House of Representatives, 1966). Of this amount, the expenditure of $5.6 million resulted in the death of 90,236 coyotes (among other losses). How many of these animals would have been taken without federal cost, how many were doing no damage, and how many were beneficial is not known.

Michigan and Minnesota were the states paying most for bounties in 1964: $254,000 and $230,000, respectively (Laun, 1965).

Also, the cost–benefit distribution is askew where most persons taking foxes do so for sport, control, or by accident, and collect bounty payment as an afterthought. In Wisconsin in 1962–1963, only 68 trappers took 20 or more foxes. Trapping a few foxes is hardly effective, yet in 1962–63, $51,927 was paid to those taking four or less foxes. Those catching 5 to 50 or more were paid only $29,829, and less than a thousand trappers or hunters were involved. It is obvious that only where trapping is thorough and sustained can there be any effective control.

Adequate predator control can be exerted briefly on a restricted area if enough removal pressure can be brought to bear. In areas where this has been accomplished, the notion arises that the same can be accomplished everywhere. Theoretically it could, but there are not enough skillful trappers to effect adequate control for large geographic areas, particularly for foxes east of the Great Plains. Yet it is on this basic erroneous assumption that bounties are paid. The good trapper catches about the same number of predators on the same area every year, until natural regulation interferes.

If bounty payments were unrealistically high, gross measurable control might be realized. So far, however, the bounty system has been used as a kind of rural dole and may be socially and politically acceptable on that basis, but the biological benefits are limited at best.

THE EXTENSION PREDATOR-CONTROL PROGRAM

Under the extension predator-control program, a qualified trapper instructs landowners and others in the art of catching destructive predators. This allows those affected to trap when and where control is needed. The educational process includes field demonstrations, information about predators, and an assessment of traps and techniques. Movies and other visual aids are used to clarify the relationship of the predator to other wildlife and to domestic livestock. It is difficult to determine the genesis of the extension predator-trapper program. The U.S. Fish and Wildlife Service and a number of states, including New York and Indiana, experimented with the concept, and in 1945 Missouri developed and promoted the system on a large scale. The first eight years of the Missouri program were reviewed by Sampson and Brohn (1955), who summarized the effort as follows:

1. Compared to other methods tried, extension predator control is the best system of providing sufficiently prompt and immediately effective service to all requests for assistance in reducing farm predator damage. Also, once farmers are trained, they can supply control measures at any time needed.

2. Simplified trapping techniques designed for busy farmers and demonstrated by able extension trappers have contributed importantly to the effectiveness of the program.

3. During an eight-year period, 635 demonstrational meetings were re-

quested and conducted throughout the State, arranged usually through county agricultural extension agents. These meetings have been attended by nearly 16,000 farmers, and 5,800 of these have participated in field training, with 873 given further individual assistance.

4. From annual mail surveys during the eight-year period, more than 1,100 farmers reported an average of 49.7 hours per year spent in trapping an average of 9.2 predators each. The total reported catch for fiscal years 1946–53 was 10,195 predators (3,099 coyotes, 5,444 red foxes, 1,555 gray foxes, 66 bobcats, and 31 wild dogs).

5. Average annual damage reduction of 81 percent, amounting to $100.37 per farmer (816 reporting) has been attained over a six-year period by farmers who applied their training. They indicate that they are well satisfied with this result.

A more recent report (Nagel, 1965) shows that predator losses in Missouri declined 35 percent since the start of the extension trapper system 20 years earlier.

This system has been advocated by the Leopold Report, the Defenders of Wildlife, the Dingell Bill (H.R. 4159), and some state conservation departments, but it has several undesirable aspects. Unskilled trappers are apt to be nonselective and inhumane in the use of steel traps. If the use of lethal chemicals is included, nonselectivity and human safety are involved.

The extension trapper program is likely to work best in the eastern half of the country where individual land ownership units are small and can be looked after directly by the owner and his family.

One of the difficulties in assessing the role of the "landowner trapper" is that state and county bounty schemes often operate concurrently. In Missouri (Nagel, Sampson, and Brohn, 1955) there was no way to determine whether the local trappers trained in the extension program used their newly acquired skill to trap for bounty, protection, or sport.

The extension trapper training program has not enjoyed equal success in all states. In Wisconsin, for example, the program was given wide publicity, and two men were employed as instructors. After a reasonably enthusiastic start, the interest bogged down and the action program collapsed. The men are now available for teaching and demonstration, but are rarely called upon. The need will vary from place to place and from time to time. The basic idea, however, is sound, and, as in Wisconsin, stand-by instructors can be given other duties when predator trapping instruction is not requested.

Regardless of the scheme used to regulate animals in pest situations, the federal government should have the major role in control programs so that uniformity of procedure, appraisal, and adjustment could be achieved.

HUNTING PREDATORS FOR SPORT

The technique of calling predators with various devices has become a challenge
to sportsmen and has brought a new dimension to sport hunting. To call a fox
to the gun by means of deceptive sounds requires enough skill to make the
effort exciting. Foxes and coyotes can be taken in this manner by the average
sportsman, and recreation is provided without seriously reducing the population of the hunted.

As an example of the growth of this sport, South Dakota began to sell non-
resident predator license stamps in 1965. In that year 121 stamps were sold,
increasing to 544 in 1966 and 549 in 1967. The Empire Fair at Sioux Falls,
South Dakota, conducts an annual predator-calling contest.

In 1966, the state of Oklahoma (Ellis, 1968) estimated that they had
4,000 sportsmen calling coyotes, bobcats, foxes, and raccoons. In addition
there were about 2,000 hunters hunting bobcats with trail hounds and about
1,500 of this number hunting raccoons in the same manner. It was estimated
that hunting for predators provided 1,372,600 man-days of sport per year, and
that coyote hunting alone accounted for 1,330,000 man-days. Hence, the
coyote is both a predator and a game animal. Other states are experiencing
the same rise in numbers of sportsmen pursuing such predators as bobcats,
foxes, coyotes, and raccoons.

Foxes and coyotes are hunted with hounds in many states. Recently, the
use of spotter light aircraft to aid mobile dogless hunter groups has become a
weekend sport activity. Yesterday's predators are today's game animals, as the
trend to predator hunting for recreation increases.

DEPREDATION INSURANCE–RARE SPECIES SUBSIDY

The Rare and Endangered Species Act, approved by the U.S. Congress on
October 15, 1966, authorizes and directs the Secretary of the Interior to consult with the states where rare and endangered species may occur; further to
consult with all interested persons and organizations to determine whether a
species is in this category; and to encourage them to take steps needed to prevent the further destruction of endangered species.

The rare and endangered predatory mammals that have thus far been listed
in the Federal Register as being threatened with extinction are the timber
wolf, red wolf, grizzly bear, Florida panther, and the black-footed ferret.

In areas where predators destroy valuable livestock (particularly sheep) one
might assume that private insurance firms would have worked out a plan for
financial protection of stockmen. However, there are few reliable actuarial

statistics on predators, and livestock data are not free from bias. In addition, difficult-to-measure variables have caused variation in the available data as to time and place of predation. It is easy to understand why insurance companies would be unwilling to take the financial risk when such data are imprecise or lacking.

The same difficulties surrounded the problem of waterfowl depredation on grain harvested on the Canadian prairies. Since no private concerns were interested, the Saskatchewan government financed an insurance scheme that has been working successfully for 13 years (see Chapter 4).

An insurance program could be developed with some of the money now used for predator control and for price supports to the sheep industry and from private donations of individuals and livestock associations. Premium payments could take into account high- and low-risk areas, use of herders, lambing paddocks, livestock protection devices, flock size, etc. The field knowledge and skills of the Wildlife Services Division field staff could be used in making loss adjustments. Any profits could be returned to the users of the insurance program.

An insurance plan does not imply that predator control can be eliminated entirely, but it could lead to greater discrimination in poisoning programs and greater effort to protect sheep against predators, since such efforts would result in lowered premium payments.

Programs for financial adjustment of damage caused by various species of wildlife now exist among the states, but there is no uniformity in the procedures used. Since it is the state's responsibility to hold the resident wildlife in trust, it is the state's prerogative to institute a depredation insurance program. Similarly, damage adjustments for migratory wildlife and rare and endangered species might well rest with the federal government.

Through insurance protection; specific, limited predator control; and more careful animal husbandry; domestic and wildlife species can live together compatibly on the same range.

Certain of the large predators are uncommon-to-rare in North America, and only the individual puma and grizzly bear occasionally prey on livestock. As of 1965, only Arizona has paid a bounty ($75) for puma. No state has a bounty on grizzly bears.

Not all predators attack livestock; hence control should be exercised against individuals and not entire species. This attitude should apply to all predators, but the situation for the rarer species is most acute.

We feel it would be wise for the federal government, as part of its program to protect rare and endangered species, to accept responsibility for the payment of predation losses incurred by private stock ranchers. Individual predators associated with bona fide losses could be taken by government trappers. Public resentment toward a species, to the extent that it is prompted by financial loss, would be alleviated.

APPENDIX: COMMON AND SCIENTIFIC NAMES OF MAMMALS

Common Name	Scientific Name	Common Name	Scientific Name
Badger	*Taxidea taxus*	Mountain lion, puma	*Felis concolor*
Bear (black)	*Euarctos americanus*	Ocelot	*Felis pardalis*
Bear (grizzly)	*Ursus horribilis*	Raccoon	*Procyon lotor*
Bison	*Bison bison*	River otter	*Lutra canadensis*
Bobcat	*Lynx rufus*	Skunk	*Mephitis mephitis*
Coyote	*Canis latrans*	Weasel	*Mustela erminea*
Fox (gray)	*Urocyon cinereoargenteus*	Wolf (gray)	*Canis lupus*
Jaguar	*Felis onca*	Wolf (red)	*Canis niger*
Mink	*Mustela vison*	Wolverine	*Gulo luscus*

Source: Jackson, 1961, and Palmer, 1954.

REFERENCES

Allen, J. A. 1876. Former range of New Zealand mammals. The Amer. Natur. 10:708–709.

Arnold, D. A. 1956. Red foxes of Michigan. Michigan Dept. of Conserv., Lansing. 48 pp.

Colorado Wool Growers Association. 1967. The sheep population continues to decrease. The Colorado Sheepman. March 1967. Mimeo. 2 pp.

Douglass, D. A., and A. M. Stebler. 1946. Bounties don't work out as they are supposed to. Michigan Conserv. 15(2):6–7, 10.

Errington, P. 1967. Of predation and life. Iowa State Univ. Press, Ames. 277 pp.

Etter, Alfred G. 1968. Inside the control empire. Installment II—The great chain of deception. Defenders of Wildlife News, 43(3):301–309.

Gottschalk, J. S. 1967. Man and wildlife (a policy for annual damage control). U.S. Dept. of Interior, Washington, D.C. 12 pp.

Jackson, H. H. T. 1961. Mammals of Wisconsin. Univ. of Wisconsin Press, Madison. 504 pp.

Laun, H. C. 1965. The bounty situation 1965. Defenders of Wildlife News 40(3):20–25.

Leopold, Aldo. 1933. Game management. Charles Scribner's Sons, N.Y. 481 pp.

Leopold, A. S. (Committee Chairman). 1964. Predator and rodent control in the United States (The Leopold Report). Trans. N. Amer. Wildl. and Natur. Resour. Conf. 29:27–49.

Nagel, W. O., Frank W. Sampson, and Allen Brohn. 1955. Predator control—why and how. Missouri Conservation Commission. 32 pp.

Nagel, W. O. 1965. Common sense in predator control. Defenders of Wildlife News, vol. 40, no. 5.

Palmer, R. S. 1954. The mammal guide. Doubleday and Co., Inc., Garden City, N.Y. 384 pp.

Sampson, Frank W., and Allen Brohn. 1955. Missouri's program of extension predator control. Journal of Wildl. Manage. 19(2):272–280.

Shapton, W. W. 1951. Do you know—. Michigan Conservation 20(1):21.

Wisconsin Conservation Department. 1964. An alternative to the fox bounty system. Report. Mimeo. 11 pp.

U.S. Department of Agriculture. 1967. Agricultural statistics (data through 1965). U.S. Dept. of Agriculture, Washington, D.C.

U.S. House of Representatives. 1966. Predatory mammals. Hearings before the subcommittee on fisheries and wildlife conservation of the committee on merchant marine and fisheries. 89th Congress, 2nd Session. U.S. Govt. Printing Office, Washington, D.C. 255 pp.

Young, S. P. 1946. The wolf in North American history. The Caxton Printers, Ltd., Caldwell, Idaho. 149 pp.

Young, S. P., and H. H. T. Jackson. 1951. The clever coyote. The Stackpole Co., Harrisburg, Pa., and the Wildlife Management Institute, Washington, D.C. 411 pp.

CHAPTER 7

Pest Situations Involving Big Game

INTRODUCTION

The term *big game* is an elastic one; it will be used here to refer to large wild herbivores, usually hoofed. There are many species of big game animals, and usually their populations constitute a valuable resource. Occasionally, however, they conflict with man's interest. These conflicts and their resolution are the subject of this report.

BIOLOGY

The well-being of big game animals depends largely on the physical condition of the forage plants. Overuse of these plants is common on sites of low productivity unless the numbers of big game animals are controlled, or the forage plants are protected by snow, periodic flooding, etc. On sites of high productivity, where forage plants are subclimax herbs or shrubs and plant succession is very rapid, big game animals tend not to overuse the range. There the food plants may dwindle and disappear because of competitive replacement by non-food plants. Also, with increased shade, the nutrient level in forage plants declines, causing malnutrition or, more properly, undernutrition, which is brought about by the shortage of specific dietary essentials (ordinarily nitrogen and phosphorus for big game). This limits population growth by lowering reproduction and increasing mortality. Big game populations may therefore be limited by situations where the forage declines in food quality, even though the plants themselves do not disappear.

Big game animals are ordinarily gregarious, often forming herds. No internal

127

mechanism of population regulation seems strong enough to keep population levels automatically in balance with environment (i.e., at or below carrying capacity). Chief among external population controls are undernutrition, predation (including hunting), epidemic disease, and such physical barriers as deep snow, floods, fences, etc.

In aboriginal America the two big game species that adapted to life in climax biotic communities were abundant. They were the bison in the prairie or steppe, and caribou in the tundra–taiga complex. Those species highly specialized for narrow ecologic niches were much less abundant. These were wild sheep, requiring rough escape country and grassland of high nutrient quality; mountain goats, requiring cliffs and high subalpine meadow forage; musk-oxen restricted to the arctic tundra; collared peccary, restricted to the arid subtropical scrub of the deep Southwest; dwarf wapiti, restricted to the overflow lands of the California interior valleys and their environs; and pronghorn, restricted to arid shrubby plains.

A third group of big game animals occupied ecotone situations and had a local abundance fluctuating with the occurrence of fire, flood, and hurricane, which created more edge or subclimax habitat. This group included white-tailed deer, mule deer, wapiti, and moose. This last group has been provided with a major expansion of habitat by such widespread occurrences as forest fires, timber cutting, and heavy grazing (plus fire suppression), causing shrub invasion in former grassland. Rigorous enforcement of hunting regulations has permitted the numbers of these species to increase greatly. Most pest situations involving big game in North America concern one of the species in this last group.

In many other parts of the world big game has been so reduced by heavy hunting that conflicts with man do not exist. But certain species, for special reasons, do become involved in pest situations. The wild boar, for example, is not much hunted in Moslem countries, because of religious injunctions against the use of swine for human food: local boar overpopulations result. In Africa, despite extensive local overkilling, pest situations involving elephants are widespread, and those involving the hippopotamus and various antelopes are often locally severe.

Because of their use of plant food, big game animals have the potential to damage agriculture wherever their populations become excessively high or wherever agriculture is pushed into regions where big game animals are abundant.

CUSTODIANSHIP AND LEGALITY

Legally, in the United States, wild animals belong to the citizens of the particular state in which they are found. The primary responsibility for the conservation or management of big game is therefore a function of state government. Ordinarily, the state legislature delegates the regulatory authority to a state department of fish and game, or of conservation, or a similar agency, which operates within policies established by an appointed commission. The responsibilities of the commission–department include alleviation of pest situations involving big game.

Certain lands within the United States are under the custodianship of federal agencies. These lands include national wildlife refuges, national forests, national parks, etc. In each case the custodial agency is authorized by law to manage, protect, or conserve the resources on those lands. This authority extends to the control of pest populations of big game. It is based on the "commerce clause" (Art. 1, Sec. 8) of the Constitution, under which the federal agencies that manage land are charged with maintenance of watershed quality, so that the navigability of downstream waters will not be impaired.

Any landowner who suffers damage from pest populations of big game may apply for redress or protection from the proper state authorities. If they do not respond to this application and the damage to private property continues, the landowner can take steps to protect his property (including killing the offending animals).

SOCIAL, POLITICAL, AND ECONOMIC ASPECTS

The United States has a peculiar system for financing big game conservation. The main financial support for state departments of fish and game comes from the sale of hunting and fishing licenses, plus taxes on hunting and sport-fishing equipment. Therefore, sportsmen have considerable political power over state wildlife management programs. This system strengthens the cultural tradition in North America, which favors the perpetuation of free hunting on all land. It also ensures that attempts to alleviate pest situations in the United States practically always start with reducing big game populations through recreational hunting, except in national parks, etc., where recreational hunting is illegal.

In contrast to the United States, most other countries vest ownership of wild animals either in the nation, or in the landowner. Responsibility for management is generally federal, with financial support through federal appropriations. Pest situations involving big game are therefore ordinarily handled by federal agencies, either directly or working through groups that control hunting rights.

HUMAN HEALTH AND SAFETY

Transmission of disease from big game to humans can occur. Big game animals constitute a reservoir of potential disease or parasitism to humans in the following situations: human handling of a freshly killed animal; human ingestion of raw or poorly-cooked animal tissue; contamination of human food with the pathogen; and human infection by blood-sucking arthropod vectors.

Leptospirosis (caused by the bacteria *Leptospira* spp.) occurs widely among wild mammals, including many species of big game. It may be contracted by humans from direct contact with the urine or organs of carrier animals. Just as butchers and slaughterhouse and meat-packing personnel become infected from domestic animals (Van der Hoeden, 1964), so hunters or biological investigators can become infected from wild animals. The possibility of such infection can be minimized by protecting the hands with rubber gloves, or by careful washing after contact.

Brucellosis (causative bacterium: *Brucella abortus*) occurs in such big game animals of temperate zones as bison, chamois, and moose (Van der Hoeden, 1964). It does not appear to be contracted from contact as readily as leptospirosis, but the possibility is there, and the same precautions are indicated. Much the same may be said for tularemia (causative bacterium: *Pasteurella tularensis*), which has been found in such big game animals as "deer," wild boar, and bear (Witenberg, 1964).

Trichinelliasis (causative parasite: *Trichinella spiralis*) occurs in big game animals that sometimes eat meat, notably the wild boar. Other animals, which do not quite fit our definition of big game but are big game animals of a sort, such as bears, seals, and walruses, also serve as a reservoir of this parasite. Man contracts it by eating the poorly-cooked meat of an infected animal. Thorough cooking destroys the organism. According to Witenberg (1964):

Special conditions for the maintenance of trichinella exist in the arctic regions, where the greater part of the fauna consists of carnivorous animals and hunting is a prevalent occupation of the population. The flesh of the animals is widely used for human consumption, while the offal, as well as the carcasses of inedible animals, are fed to sledge dogs or thrown away. This situation results in a high incidence of trichinelliasis in men and animals.

Echinococcosis or hydatid disease (causative parasite: *Echinococcus granulosus*) is ordinarily found in humans as a result of contamination of human food by feces of the domestic dog that contain *Echinococcus* eggs. The dogs become infected by ingesting the intermediate stage of this tapeworm when they eat such big game animals as caribou and moose. It is most common in the Far North. ". . . on the Canadian mainland the hydatids occur in man mostly and in animals exclusively in the lungs; (b) they are found in the larger deer (caribou, moose, wapiti), but not in other wild or domestic herbivores . . . (and are) mainly propagated in the sylvatic life cycle (i.e. deer-wolf-deer with a side branch to dog-man)" (Witenberg, 1964:665–667). This disease can be avoided by the regular medical treatment of dogs, or by adequate care in food preparation.

Human trypanosomiasis (causative parasites: *Trypanosoma* spp.) occurs in

Africa in a zone of about 15° north and south of the equator, where its insect vector, the tsetse fly (*Glossina* spp.), is found. Van Riel (1964) says:

In essence, African trypanosomiasis of man presents two epidemiological aspects: one in areas where the animal (antelope) is an important reservoir of the parasite (*T. rhodesiense*) and where the zoophilic glossinae are the vectors; the other where the transmitting flies live in closer contact with man who is there practically the sole reservoir of the trypanosome (*T. gambiense*).

Recommendations for the control of human trypanosomiasis of the first sort center on reducing the possibility of contact between the tsetse fly and man, regrouping the scattered human population, and taking ecological measures to reduce the fly (Van Riel, 1964).

HUMAN SAFETY AND BIG GAME

Big game threaten human safety in two ways: by direct attacks, and by collisions between vehicles and game.

Direct attacks are few and serious results are still more rare. Undoubtedly the commonest attack situation involves a pet animal that has matured and entered breeding condition and suddenly becomes antagonistic. (Danger to man from large carnivores and bears is covered in another section of this report).

Collisions between vehicles and game are a much more common and serious matter. Severe damage to the driver and the vehicle as well as to the big game often result. This is a situation typical of highly industrialized countries with good roads and much fast traffic and where effective protection of big game animals has permitted large populations to build up. The species most commonly involved are roe deer in Europe and white-tailed deer in North America. The most dangerous situation is where a high-speed highway runs through heavy vegetation harboring high populations of big game animals. Most collisions occur at night.

Traffic is increasing, both in quantity and in speed, and deer populations are also increasing in many areas. These trends are reflected in the results of nationwide surveys made by F. A. Thompson. The following table gives the estimated financial losses from deer–automobile collisions for 1963–1966, based on data from Thompson (1966) and on communications from various state fish and game departments:

Year	Number of Collisions	Financial Loss
1963	84,950	$16,990,000
1964	90,650	21,846,650
1965	109,300	31,369,000
1966	119,198	32,500,000

White-tailed deer were most often involved in the collisions, and the incidence was highest in May and November. Actual losses are probably higher than the estimates given, since collisions tend to be under-reported, especially when damage is minor.

Human injuries often result from these collisions. If we assume one human injury for each 40 reported collisions, minimal estimates for human injury were 2,120 in 1963 and 3,000 in 1966. Fortunately, few deaths result.

The main attempts to alleviate this situation have been the erection of warning signs, reduction of roadside cover, fencing of the highway, erection of mirrors along the roadside, and the reduction of the big game population through recreational hunting. These have been effective, but an acceptable level of success has not yet been attained.

Another situation leading to collisions is when there is heavy snow and the highway is plowed free. Big game animals, most frequently moose, get into the cleared path, travel along it, and are unable or reluctant to scale the bordering snowbank when a vehicle approaches.

HUMAN POISONING AND BIG GAME

There are two patterns of human poisoning by ingestion of big game meat: the passage of radioactive contaminants from caribou and the passage of DDT from various big game species.

In the Arctic there is an extremely heavy accumulation of radionuclides, especially on lichens (which do not shed their "leaves"). These lichens form an important food of caribou, which consequently ingest the radionuclides and become correspondingly "hot." To date, no one has demonstrated whether or not the caribou are harmed by this. Humans eat caribou and may thereafter emit radiation. The public health significance of this is not yet known (Pruitt, 1963; Hanson, Whicker, and Dall, 1963; Hanson *et al.*, 1963; Davis, Hanson, and Watson, 1961).

Similarly, there is a potential threat to human health through ingestion of DDT or other pesticides (or their breakdown products) through eating big game animals. Buckley (1963) states, "Deer collected from an area sprayed with 1 pound per acre of DDT for spruce budworm control contained DDT residues, some in excess of legal tolerance for domestic meats. . . . One may well question whether it is rational to eat game meats containing residues that would be unacceptable in domestic foods. There seem to be three possible solutions to this problem: (1) reduction in use of persistent toxic pesticides to the point that unacceptable residues no longer occur; (2) closing of seasons to prevent the taking of game known to contain unacceptable residues . . . ; or (3) permitting the taking, but discouraging the eating, of game."

BIG GAME AND AGRICULTURAL LIVESTOCK

CONFLICT FOR FORAGE

Food habits of some big game and of domestic livestock species overlap, and competition may occur. For competition to be real, the following conditions must be met: the two species of ungulate must eat the same plant; the feeding areas must be the same; and at least some of the plants shared must be in short supply.

When competition reaches a level at which the livestock operator suffers significant economic loss, a pest situation exists. Principal examples of this sort of competition between big game and livestock are elk/cattle in the northern Rockies (Morris, 1956), mule deer/sheep in the central Rockies and Great Basin (Smith and Julander, 1953), and white-tailed deer/cattle or (in Texas) sheep and goats (Grelen and Thomas, 1957). In the last case a quantitative assessment was made in terms of forage consumption, and six adult deer were equated with six adult sheep or goats or one mature cow (Merrill, 1957).

The direct solution for such situations is reduction of either big game or of livestock, or both. The social, political, or economic problems involved are often severe.

BIG GAME AS RESERVOIRS FOR LIVESTOCK PARASITES

A complete review of the role of big game animals as reservoirs for livestock diseases and parasites is beyond the scope of this report. The most thoroughly studied North American game animal from this aspect is the white-tailed deer. Its diseases and parasites have been reviewed by several authors, the most comprehensive works being those of Olson and Fenstermacher (1943), and more recently, Anderson (1962a,b). Of the 66 or more known parasites of white-tailed deer, at least 36 are transmissible to domestic livestock; many, such as liver fluke and lungworms, can be highly pathogenic. The role of white-tailed deer as reservoirs of the American liver fluke (*Fascioloides magna*) has been discussed by Olson (1949) and Griffiths (1962). More recently, this deer has been implicated as the carrier of the parasite *Pneumostrongylus tennuis*, which does not greatly affect the health of the deer but is often fatal to moose. The northward spread of the white-tailed deer into moose range may thus result in serious consequences for the latter (Anderson, 1965).

Cowan (1946) has contributed an extensive review of the diseases and parasites of the Columbian black-tailed deer. Longhurst and Douglas (1953) discuss the interrelationships of these deer and domestic sheep. The internal parasites of mule deer were surveyed by Landram (1951) and Landram and Honess (1955). Good review papers on the parasites of antelope include Goldsby and Eveleth (1954) and Gilmore and Allen (1960). Alderson (1949) and Cowan (1951) studied the parasites of elk, probably the least known of the North American game animals in this regard.

Investigations of the parasites of moose were carried out by Wallace (1934) and Olson and Fenstermacher (1942). Moose are probably the most important game reservoir for hydatid disease, the intermediate stage of the tapeworm *Echinococcus granulosus.* This parasite is highly pathogenic and infective to both man and livestock. The role of moose and other game animals as a reservoir for this tapeworm is discussed by Harper, Ruttan, and Benson (1955).

In general, the transmission of parasites from game to livestock or to other game species is directly related to the amount of deposition of infective stages on the range. This, in turn, is related to the size of the game population and the percentage of parasitism. Often overpopulation and severe parasitism go hand in hand. Depletion of range causes malnutrition in game herds, making them more susceptible to parasitism and thus increasing the amount of contamination in their droppings. Anderson (1962a) emphasizes that maintaining healthy game populations and eradicating parasites in livestock can minimize such pest situations.

BIG GAME AS RESERVOIRS FOR OTHER DISEASES OF LIVESTOCK

Comprehensive papers on disease interrelationships among wildlife and domestic livestock include those of Shillinger (1937, 1942) and of Herman (1945). There are many diseases with bacterial, rickettsial, or viral pathogens that have been reported for most domestic and many wild ungulates. They may be quite serious for the infected animal. But for a pest situation to exist, the wild population must provide a reservoir from which domestic animals are readily and repeatedly infected. There was a short-lived episode in 1924–1926 when hoof-and-mouth disease (of which the pathogen is *Aphthae epizooticae*) was introduced to California through infected cattle. Deer could have been infected, and an attempt was made to eliminate all the deer in the critical locality. The disease did not become established (Leopold *et al.*, 1951). No such situation now exists, apparently, in North America.

Bovine trypanosomiasis, like the human trypanosomiasis mentioned earlier, is carried by some big game animals of equatorial Africa. The disease, called *nagana*, is transmitted to domestic livestock by the tsetse fly (*Glossina* spp.); it makes animal breeding difficult or impossible. Van Riel (1964, p. 510–511) says:

In equines, nagana usually assumes an acute course. . . . The disease is also serious in . . . camels, swine, and dogs. Cattle, sheep and goats are definitely more resistant; the course of nagana in these species is usually chronic or even symptomless although some strains have been reported to cause severe mortality in cattle.

The use of Antrycide made chemoprophylaxis of nagana possible. This should be associated with steps for control of *Glossina*, such as bush-clearing, use of insecticides and destruction of game (though this latter measure must be carried out with discretion as otherwise it may be harmful).

Darling (1965) comments as follows:

The slaughter policy of the past achieved some clearing up of the problem and human settlement and establishment of maize farms did help to contain the tsetse. But the slaughter policy was finally ineffective because the warthog, the kudu and the duiker are precisely the species which are able to survive such a policy and infiltrate once more.

Tsetse control to be effective searches for all possible angles of attack and employs some or all the weapons it can muster at different times and in particular situations. These weapons include insecticides, land-use and habitation patterns, bush clearance, biological control through parasitism and genetic sterilization and . . . game reduction or even elimination.

A different approach to the solution of nagana losses in livestock is to avoid the introduction of livestock into nagana areas and "farm" the native ungulates for meat and hides instead (Dasmann, 1964).

DAMAGE TO AGRICULTURAL CROPS

Generally speaking, plants growing on soils that are fertile, or fertilized heavily, or irrigated in arid regions, are more attractive to herbivorous animals than plants growing elsewhere (Mitchell and Hosley, 1936). Since the more fertile soils of any region tend to be taken over for agriculture, and since the crops are often artificially fertilized as well, many crops become more attractive to big game than their natural forage.

Thus, we can distinguish two kinds of pest situations: one where the crop is more attractive to big game than their natural forage, and the other where the crop is merely one of many available foods. In the former, there will be at least local damage to crops as long as big game animals can reach them. The only solution is to eliminate the big game completely or to protect the crop.

Valuable plants may sometimes be protected by externally applied repellents, although the results vary from place to place, even with the same repellent and target species. Besser and Welch (1959) report on two deer repellents, ZAC (zinc dimethyldithiocarbamate, cyclohexylamine complex) and TMTD [bis(dimethylthiocarbamoyl) disulfide], used in orchards and on young coniferous trees. Although more critical research needs to be done, their work indicates that they seem useful in protecting the vital parts of the tree. Carpenter (1967) reports the successful use of tankage, a slaughterhouse product.

Currently, attempts are being made to develop a systemic repellent that is taken up by the roots of the plant and deposited in the parts ordinarily browsed by big game.

Crops are especially prone to damage when fields are small and near big game cover. Fencing may be a practical measure against relatively small animals. Small orchard trees can be fenced individually, or whole areas of such attractive crops as fruit trees, vines, or irrigated pastures may be fenced. New designs for overhanging fences and for electric fences make this more economical than formerly (Longhurst *et al.*, 1962; Seamans, 1951).

The expense of fencing, however, still spurs the search for cheaper methods of protecting crops from big game animals. Some measures that have worked well (at least in local instances) are the use of dogs, the nightly playing of radios within the fields, and the stringing of exploding booby-traps across access routes. These have an advantage in that the big game animal is already conditioned to be afraid of dogs, human voices, and unexpected explosions (Flyger and Thoerig, 1961, 1962).

Certain species of big game are often strongly attracted to a particular crop. Buckwheat attracts white-tailed deer; potatoes, wild boar; and meadow hay, wapiti. In such cases, fencing, elimination of the game, or a change of crops is the only remedy.

In East Africa, the rapid increase and expansion of the human population has led to many complaints of damage by elephants. A study of the elephant in Uganda (Brooks and Buss, 1962) has shown that an average of one thousand elephants have been shot annually between 1927 and 1958 by the government's Elephant Control Department. However, at the beginning of that period some 70 percent of the area was inhabited by elephants, while by 1958 this had been reduced to only about 17 percent. Yet the elephants shot annually during the first five years of record (1927–1931) averaged 879 and those shot during the last five years (1954–1958) averaged 851, an increase in intensity of control-killing per unit of area of over 300 percent, an obvious parallel to the increase of the human population of over 50 percent during the same period. Much the same pattern of conflict between elephant and man has developed in other African countries.

DAMAGE TO FORESTS

Where the crop damaged by big game is merely one of a number of foods, the amount of damage to the crop is related to the population density of big game and the quality of the habitat (often called the *carrying capacity*) for the species of big game in question. Some findings of Ueckermann (1960) for red deer (*Cervus elaphus*) illustrate these points. Red deer damage Norway spruce and beech trees by stripping the bark. Generally the damage-index rises as the red-deer density rises, but on good range a higher density of red deer can be maintained with less forest damage than would occur on poor range.

In North America big game (particularly deer) damage young timber reproduction, usually by nipping the terminal bud. In extreme cases, browsing kills the tree, and certain species are practically eliminated from the forest plant association. For example, in the hemlock-hardwood forest of the Great Lakes region, white-tailed deer in excessive densities will practically wipe out reproduction of hemlock, yellow birch, and white cedar (Graham, 1954). The relation of white-tailed deer density and damage to mixed conifer–hardwood forests in North America is given in Table 1.

In the management of coniferous forests, it is usual to clear-cut blocks of mature timber, and then obtain even-aged reproduction. In western North America, the black-tailed race of the mule deer causes damage to redwood, Douglas fir, and ponderosa pine seedlings, especially by nipping the terminal buds.

An intensive study (Hines, 1963) in the coastal Douglas fir forest of Oregon showed the following relation between black-tailed deer density and damage to Douglas-fir seedlings:

Deer Density (No./sq mi)	Seedlings Damaged (%)
28	2
65	14–20
81	45

There are only a few estimates of the actual economic losses to commercial crops caused by big game animals. In Pennsylvania farmland Pasto and Thomas (1954 and 1955) found that where deer damage to crops was obvious, it caused an annual average loss of $26.00 per acre. Annual losses caused by white-tailed deer to commercial forests in Pennsylvania have been estimated at $0.87/acre with light deer populations and up to $23.19/acre with heavy deer populations (McCullough, quoted by Bennett, 1962). In one ponderosa pine forest in Montana, a heavy white-tailed deer population was estimated to cause an annual loss of $2.80/acre (Neils, Adams, and Blair, 1955).

CONTROL

Where other measures are ineffective, the only solution to crop damage by big game is a combination of direct reduction and harassment. If control is concentrated in the areas where damage is greatest, it will have the double effect of removing some animals of the pest population, and frightening others away.

TABLE 1 Relation of White-Tailed Deer Density to Damage in Some North American Conifer–Hardwood Forests

Estimated Deer Density (No./Sq Mi)	Damage to Timber Reproduction	Reference
110	Total	Bartlett, 1958
35–45	Severe	Stoeckler, Strothman, and Krefting, 1957
25–50	Severe	Graham, 1954
20+	Moderate	Bennett, 1962
20	Light	Bartlett, 1958
8–12	Light	Graham, 1954
8–9	Light	Bennett, 1962

To be efficient in such an undertaking, it is important to know enough about the movements of the pest population to be able to bring pressure on it, and not just on the species at large.

Where a build-up of a big game population beyond a certain level causes the development of a pest situation (damage to agricultural crops, forest regeneration, watershed cover, etc.), then an effort must be made to hold the general population to the proper level. Overpopulations tend to develop in areas where protective legislation is particularly effective, i.e., in highly industrialized nations; where the country is particularly difficult to penetrate, as in New Zealand; where the big game species involved is protected by religious injunctions, as the wild boar in Moslem areas or the cow-like big game animals in India; or where the animal is especially large, strong, and wide-ranging, as the elephant in East Africa.

A current example of big game damage to watershed and other vegetation is found in New Zealand, where many mammals have been introduced by man. Howard (1964) lists damage by feral goat, chamois, red deer, and feral pig as acute, and that by tahr, Japanese deer, and fallow deer as moderate. After a series of increasingly strong efforts to control big game animals causing damage to forest and watershed cover, the Noxious Animals Act was passed in 1956, which permitted the New Zealand government wide powers in animal control. Shooting and poisoning are the main methods of local control; as yet, widespread control has been possible only in open country (Howard, 1964).

The wild boar in West Pakistan constitutes a serious threat to agricultural crops. Strong efforts to control boar populations have included shooting, poisoning, and most recently, the introduction of a disease, hog cholera. So far, these measures have been only locally effective.

In the United States, when the hunting pressure required to control a big game population is light or moderate, and where legislation and public opinion permit this pressure to be applied, the major part of the control is by recreational hunters. The measure of success in this approach varies according to the species, particularly with regard to how easy it is to hunt and how attractive it is to the sportsman. The pronghorn, at one extreme, is quite easy to control by the manipulation of hunting pressure. The wapiti, a popular game animal, is kept almost stable by hunting. On the other hand, to stabilize the mule deer would require an estimated increase of 200 percent in the present annual kill (Taber, 1966).

A number of countries have adopted a system of hunting administration based on leases to hunter groups. Excessive populations of big game animals on a given area may then be controlled by that particular group. This method is being applied against the wild boar in Turkey. In East Africa, professionals generally control big game animals that damage agricultural crops; in fact, animal control is a major activity of many game departments in that region.

Payments for game damage are made by a few states in the United States. This provides at least a crude monetary measure of the magnitude of agricultural damage by big game in those states. Wisconsin, for example, in 1965–1966 had 310 claims of agricultural damage by white-tailed deer, totaling $62,728 (Wisconsin Conservation Department, 1966). In most states, however, payments from the state for agricultural damage are regarded as prohibitively expensive to investigate and pay.

A special sort of pest situation exists in a number of countries in national parks and other sanctuaries where recreational hunting is forbidden by law. Here big game populations tend to increase and ultimately to damage their own food supply, watershed cover, and habitat for other animals. The wapiti population in Yellowstone National Park is a case in point. In 1963, a policy covering such pest situations was proposed by a committee appointed by the U.S. Secretary of the Interior and was accepted for application in the national parks. According to Leopold (1963), it was recommended that

> Where other methods of control are inapplicable or impractical, excess park ungulates (hoofed animals such as deer, elk, mountain sheep) must be removed by killing. It is the unanimous recommendation of this Board that such shooting be conducted by competent personnel, under the sole jurisdiction of the National Park Service, and for the sole purpose of animal removal, not recreational hunting.

Where it is necessary to reduce big game numbers by direct control, the control operations attempt to stem the natural tendency of the population to increase by reproduction. Research is now being conducted on chemical inhibitors of reproduction, but as yet there are no practical applications to pest situations involving big game. Another approach is to shift the sex ratio in the big game population by directing control measures principally against females.

CONCLUSION

Although pest situations involving big game animals are largely undocumented from an economic point of view, they can be of great local importance. They have tended to increase as big game populations have increased due to protection, and as human use of the landscape for other purposes has become more intensive. The size of the offending animals makes their control by physical means expensive, and their obvious desirability for sport and aesthetics makes efforts to reduce offending populations by killing unpopular with the general public.

The pest situations most difficult to resolve, at present, appear to be local damage to agricultural and forest crops; damage to vehicles due to collisions with big game animals; range damage due to big game overabundance in parks and sanctuaries; and the functioning of big game populations as reservoirs of disease or parasitism transmissible to other big game species, livestock, or human beings.

In each of these situations the alleviation of the problem requires (a) the identification of the actual subpopulation involved and (b) the application of control measures to that population. Improved methods need to be developed for each of these processes.

APPENDIX: COMMON AND SCIENTIFIC NAMES OF BIG GAME ANIMALS

Common Name	Scientific Name	Common Name	Scientific Name
Bison	*Bison bison*	Musk-ox	*Ovibos moschatus*
Caribou	*Rangifer tarandus, R. caribou, R. arcticus*	Polar bear	*Thalarctos maritimus*
		Pronghorn	*Antilocapra americana*
Chamois	*Rupicapra rupicapra*	Red deer	*Cervus elaphus*
Collared peccary	*Pecari angulatus*	Roe deer	*Capreolus capreolus*
Duiker	Various species of *Cephalophus* and *Sylvicapra*	Seal	Various species of *Phoca, Erignathus, Halichoerus,* and *Cystophora*
Dwarf wapiti	*Cervus nannodes*	Tahr	*Hemitragus jemlahicus*
Fallow deer	*Dama dama*	Walrus	*Odobenus rosmarus*
Japanese deer	*Cervus nippon*	Wapiti	*Cervus canadensis*
Kudu	*Tragelaphus* species	Warthog	*Phacochoerus aethiopicus*
Moose	*Alces alces*	White-tailed deer	*Odocoileus virginianus*
Mountain goat	*Oreamnos americana*	Wild boar	*Sus scrofa*
Mule deer	*Odocoileus hemionus*	Wild sheep	*Ovis canadensis* and *O. dalli*

Source: Walker *et al*, 1964

REFERENCES

Alderson, L. E. 1949. Internal parasites of the elk in Wyoming (Abstr. of thesis). University of Wyoming Publ. 16:77–78.

Anderson, R. C. 1962a. The parasites (helminths and arthropods) of white-tailed deer, Proc. 1st Natl. White-tailed Deer Symposium. Pp. 162–173.

Anderson, R. C. 1962b. The helminth and arthropod parasites of the white-tailed deer (*Odocoileus virginianus*), a general review. Trans. Royal Canadian Inst. 34(1):57–92.

Anderson, R. C. 1965. Cerebrospinal nematodiasis (*Pneumostrongylus tenuis*) in North American cervids. Trans. N. Amer. Wildl. Conf. 30:156–166.

Bartlett, C. O. 1958. A study of some deer and forest relationships in Rondeau Provincial Park. Ontario Dept. Lands and Forests, Biol. Ser. 66. 137 pp.

Bennett, A. L. 1962. Industrial forestry and wildlife—the northeast. J. Forest. 60(1):118–120.

Besser, J. F., and J. F. Welch. 1959. Chemical repellents for the control of mammal damage to plants. Trans. N. Amer. Wildl. Conf. 24:166–173.

Brooks, A. C., and I. O. Buss. 1962. Past and present status of the elephant in Uganda. J. Wildl. Manage. 26(1):38–50.

Buckley, J. L. 1963. Effects of pesticides upon wild birds and mammals, pp. 23–29. *In* Pesticides, their use and effect, proceedings of a symposium sponsored by the New York State Joint Legislative Committee on Natural Resources, Albany.

Carpenter, M. 1967. Control of deer damage. Virginia Wildl. 23(5):8–9.

Cowan, I. McT. 1946. Parasites, diseases, injuries, and anomalies of the Columbian black-tailed deer (*Odocoileus hemionus columbianus* Richardson). Canadian J. Res. (D) 24:71–103.

Cowan, I. McT. 1951. The diseases and parasites of big game mammals of western Canada, pp. 37–64. *In* Proc. 5th Ann. Game Conf., Victoria, B.C.

Darling, F. F. 1965. Control of tsetse fly by "game" extermination. I.U.C.N. Bull. N.S. 16:7.

Dasmann, R. F. 1964. African game ranching. The Macmillan Co., N.Y. 75 pp.

Davis, J. J., W. C. Hanson, and D. G. Watson. 1961. Some effects of environmental factors upon accumulation of worldwide fallout in natural populations, pp. 35–38. *In* Radioecology: Proc. First Natl. Symposium and Radioecology. V. Schultz and A. W. Klement, Jr. (eds.). 1963. Reinhold Publishing Corp. and A.I.B.S.

Flyger, V., and T. Thoerig. 1961. Preliminary report on a new principal for prevention of crop damage by deer. Proc. Ann. Conf. S. East. Ass. Game and Fish Comm. 15:119–122.

Flyger, V., and T. Thoerig. 1962. Crop damage caused by Maryland deer. Proc. Ann. Conf. 16:45–52.

Gilmore, R. E., and R. W. Allen. 1960. Helminth parasites of pronghorn antelope (*Antilocapra americana*) in New Mexico, with new host records. Proc. Helm. Soc. Washington 27(1):69–73.

Goldsby, A. I., and D. F. Eveleth. 1954. Internal parasites in North Dakota antelope. J. Parasitol. 40(6):637–648.

Graham, S. A. 1954. Changes in northern Michigan deer forest from browsing by deer. Trans. N. Amer. Wildl. Conf. 19:526–533.

Grelen, H. E., and G. W. Thomas. 1957. Livestock and deer activities on the Edwards Plateau of Texas. J. Range Manage. 10(1):34–37.

Griffiths, H. J. 1962. Fascioloidiasis of cattle, sheep and deer in northern Minnesota. J. Amer. Vet. Med. Ass. 140:342–347.

Hanson, W. C., F. W. Whicker, and A. H. Dall. 1963. Iodine 131 in the thyroids of North American deer and caribou: comparison after nuclear tests. Science 140(358):801–802.

Hanson, W. C., A. H. Dahl, F. W. Whicker, W. M. Longhurst, V. Flyger, S. P. Davey, and K. R. Greer. 1963. Thyroidal radioiodine concentration in North American deer following 1961–1963 nuclear weapons test. Health Phys. 9:1235–1239.

Harper, T. A., R. A. Ruttan, and W. A. Benson. 1955. Hydatid disease (*Echinococcus granulosus*) in Saskatchewan big game. Trans. N. Amer. Wild. Conf. 20:198–208.

Herman, C. M. 1945. Deer management problems as related to disease and parasites of domestic range livestock. Trans. N. Amer. Wildl. Conf. 10:242–246.

Hines, W. W. 1963. Relationship of black-tailed deer density to conifer survival. Proc. West. Ass. of State Game and Fish Comm. 43:188–192.

Howard, W. E. 1964. Animal control in New Zealand. Proc. 2nd Vert. Pest Contr. Conf., pp. 117–126.

Landram, J. F. 1951. Internal parasites of the mule deer (*Odocoileus hemionus* subspecies) in Wyoming (Abstr. of thesis). University of Wyoming Publ. 16:154.

Landram, J. F., and R. F. Honess. 1955. Some internal parasites of the mule deer (*Odocoileus hemionus hemionus*) in Wyoming. Wyoming Fish and Game Comm.

Leopold, A. S., T. Riney, R. McCain, and L. Tevis, Jr. 1951. The Jawbone deer herd. California State Div. of Fish and Game, Game Bull 4. 139 pp.

Leopold, A. S. (Committee Chairman). 1963. Study of wildlife problems in national parks. Trans. N. Amer. Wildl. and Nat. Resour. Conf. 28:28–45.

Longhurst, W. M., and J. R. Douglas. 1953. Parasite interrelationships of domestic sheep and Columbian black-tailed deer. Trans. N. Amer. Wildl. Conf. 18:168–188.

Longhurst, W. M., M. B. Jones, R. R. Parks, L. W. Newbauer, and M. W. Cummings. 1962. Fences for controlling deer damage. Univ. of California, Agr. Exp. Sta. Circ. 514. 19 pp.

Merrill, L. B. (Committee Chairman). 1957. Livestock and deer ratios for Texas range lands. Texas Agr. Exp. Sta. (Coll. Sta.), MP-221. 9 pp.

Mitchell, H. L., and N. W. Hosley. 1936. Differential browsing on plots variously fertilized. Black Rock Forest Papers (New York) 1(5):24–27.

Morris, M. S. 1956. Elk and livestock competition. J. Range Manage. 9(1):11–14.

Neils, G., L. Adams, and R. M. Blair. 1955. Management of white-tailed deer and ponderosa pine. Trans. N. Amer. Wildl. Conf. 20:539–551.

Olson, O. W. 1949. White-tailed deer as a reservoir host of the large American liver fluke. Vet. Med. 44:26–30.

Olson, O. W., and R. Fenstermacher. 1942. Parasites of moose in northern Minnesota. Amer. J. Vet. Res. 3:403–408.

Olson, O. W., and R. Fenstermacher. 1943. The helminths of North American deer, with special reference to those of white-tailed deer (*Odocoileus virginianus borealis*) in Minnesota. Tech. Bull. 159, Minnesota Agric. Exp. Sta. 20 pp.

Pasto, J. K., and D. W. Thomas. 1954 and 1955. Deer economics. Pennsylvania Game News 25(11–12) and 26(1–2).

Pruitt, W. O. 1963. Lichen, caribou and high radiation in Eskimos. Audubon Mag. 65(5):284–287.

Seamans, R. 1951. Electric fences for the control of deer damage. Vermont Fish and Game Serv. 77 pp. (from Flyger and Thoerig, 1962).

Shillinger, J. E. 1937. Disease relationships of domestic stock and wildlife. Trans. N. Amer. Wildl. Conf. 2:298–302.

Shillinger, J. E. 1942. Diseases of wildlife and their relationship to domestic livestock, pp. 1217–1225. *In* Yearbook Agric., U.S. Dept. Agriculture.

Smith, J. G. and O. Julander. 1953. Deer and sheep competition in Utah. J. Wildl. Manage. 17(2):101–112.

Stoeckler, J. H., R. O. Strothmann, and L. W. Krefting. 1957. Effects of deer browsing in the northern hardwood-hemlock type in northeastern Wisconsin. J. Wildl. Manage. 21(1):75–80.

Taber, R. D. 1966. Land use and native cervid populations in America north of Mexico. VI Congr., Int. Union of Game Biologists Proc., pp. 201–225. Reprinted as Bull. 29, Montana Forest and Conservation Exp. Sta.

Thompson, F. A. 1966. Deer on highways. New Mexico Dept. of Fish and Game (two of a series of reports). Mimeo.

Ueckermann, E. 1960. Wildstandbewertschaftung und Wildschadenverhutung beim Rotwild. Verlag Paul Parey, Hamburg and Berlin. 162 pp.

Van der Hoeden, J. (ed.). 1964. Zoonoses. Elsevier Publishing Co. 774 pp.

Van Riel, J. 1964. Trypanosomiases, pp. 505–521. *In* Van der Hoeden (1964).

Walker, E. P., F. Warnick, K. I. Lange, H. E. Uible, S. E. Hamlet, M. A. Davis, P. F. Wright. 1964. Mammals and the world. 3 vols. Johns Hopkins Press, Baltimore, Vol. 1, 644 pp.; Vol. II, 855 pp.; Vol. III, 769 pp.

Wallace, F. G. 1934. Parasites collected from moose, *Alces americanus*, in northern Minnesota. J. Amer. Vet. Med. Ass. 112:770–778.

Wisconsin Conservation Department. Intradepartmental memo. Oct. 11, 1966. From N. R. Barger to J. R. Smith (1965–66 fiscal year deer and bear damage report). Mimeo.

Witenberg, G. G. 1964. Nematodiases, pp. 530–601. *In* Cestodiases [subsection d] Echinococcosis; pp. 661–682 *in* Van der Hoeden, 1964.

CHAPTER 8

Where We Stand

Man has the prerogative of controlling animals that threaten his health, economy, comfort, recreation, and, ultimately, his survival. Our advancing technology has provided the control methods but they lack adequate safeguards for nonpest species and, in some cases, for man himself. In such situations, we must make value judgments on man's confrontation with other vertebrates.

Judgments of this kind cannot rely on obvious cause-and-effect relationships. Many man-animal interactions are multifaceted. Problems are often ill-defined ecologically, legally, and socially. To understand the difficulty of effecting ecologically sound relief from pest situations, one need only be aware of the impact that habit and tradition play in scientific developments. Extrapolation of control methodology from one geographic, cultural, or political area to another is itself a value judgment.

In the United States we are able to appraise our pest problems from the viewpoint of many disciplines; in addition, we have to cope with most pest situations. Our greatest need is to systematize our efforts so that safety and efficiency can be achieved. Long-range benefits and safeguards must take precedence over expediency.

Modern man mistakenly assumes that pest situations will always be resolved in his favor. More and more, as population growth diminishes usable living space, he will be forced to live with his self-imposed pest situations, for there will be few acceptable alternatives.

Man is the only animal with the capacity to exercise conscious restraint, but it will avail him nothing unless he understands the interrelation of living things. If this responsibility is taken lightly, man may alter the natural order and indirectly threaten his own place in the new order thus created.

RESEARCH NEEDS

Although most pest control research is devoted to improving orthodox techniques, pest situations cannot be alleviated by control methodology alone. That an animal is a pest means that the environment has been altered. Hence, the prime requisite to control is a detailed knowledge of the biology of the species to be controlled and of its environment, by which the weakest link in the system can be located and assessed. Once the link at which control is likely to be most effective has been identified, control techniques can be adapted to give the methodology maximum efficiency.

When immediate control is imperative, the fundamental need for ecological information does not preclude developing a safe, if less than ideal, operational program on the basis of current knowledge. Basic research and control evaluation studies should be conducted at the same time. The resource cannot remain unprotected until research produces the ideal control, although research may eventually alter the biological frame of reference and thus render initial hypotheses academic.

Extensive knowledge is required to achieve control without adversely affecting other organisms and habitats. Knowledge gaps, even in the fundamentals of control, are great, but efforts in the following areas of investigation should substantially improve matters:

• Vulnerable points in the ecology of a species should be sought so that control methods can be applied most effectively.

• Reproduction inhibitors are among the more promising control techniques. They are humane, nonlethal, easy to use in the field, and reasonably economical. They are not secondary contaminants. Their full potential is unrealized because available materials are not specific to the species or animal, their effect is short-lived, and their efficacy and economy need to be improved.

• Research is needed on ways to avoid pest damage, particularly by adjusting behavior or altering the physiology of animals, or by adjusting human behavior and economy. If operating procedures can be changed efficiently and economically, traditional attitudes can be broken down and the pest situations eliminated.

• Research may uncover ways of uniting two relatively inefficient control systems into one that is effective. Too often unilateral control agents are used where combinations of various types would give better results.

• There is some evidence that additional research in ultrasonics and high intensity light would prove fruitful.

• Biological control of vertebrates, although it has sometimes created a secondary pest situation rivaling or exceeding the initial problem, is basically sound. Research in this area should minimize unsound programs and at the same time develop biological-control systems that can be successfully applied.

● Research in the social sciences, including jurisprudence as it relates to defining pests and pest situations and responsibilities and methods for control, is needed if effective programs are to be developed.

● Toxicants that are specific, humane, fast-acting, and safe for the user should be continuously sought. These should be used when other approaches fail.

● Research to develop accurate methods for assessing control programs may be as important as improving control *per se*, for the success of any control program depends on knowing what part the control technique or material plays.

● Specific assessment is needed of the ramifications of control (or lack of control) on local and national economy, of the economics of control itself, and of the indirect effects of control on economies of associated natural resources, for legislation should be based on an appraisal of economic as well as noneconomic values.

TRAINING FOR PEST CONTROL

Howard (1962) has long advocated training to upgrade the skills of those who must undertake pest control. Apparently, the only formulated college curriculum was presented by Jackson (1964). But his curriculum was not focused on pest control or applied biology (except for two 3-credit courses in economic biology), and it included only one course in ecology. The proposed curriculum was nonetheless well-balanced and suitable for training in a number of biological fields. Indeed, any sound curriculum in biology, when geared to ecology and accompanied by attendant subjects in chemistry, physics, and mathematics, can form the academic base for work in vertebrate management.

A basic change is needed in public attitudes toward control specialists. The public should have a better understanding of the qualifications needed for entering the field of vertebrate control. The control biologist may need even stronger academic background than the majority of field biologists.

In many respects pest problems go beyond biology. They are not always best resolved by the management of the animal in question, but rather by the modification of human behavior. Training in the social sciences is needed by anyone who undertakes such manipulation.

At the college level a special curriculum can be developed that will give students classical courses in the biological and social sciences that bear on control. Supplementing this, field assignments in applied aspects, including perhaps an internship with professionals, would be excellent at the bachelor's degree level. For more advanced training, a postgraduate program should be developed.

As an alternative to the baccalaureate degree, technical schools could provide a two-year program in the biological, chemical, and mechanical aspects of vertebrate control. The same approach might also be taken, on a less intensive scale, through short courses or special terms in vocational schools.

On-the-job training for field personnel would be desirable, but the instructors must be knowledgeable. The effectiveness of such training depends greatly on the amount of time devoted to it.

Finally, institutes and workshops for teachers or field workers could increase their comprehension of the problems and of the techniques and materials available to cope with them.

While training in techniques and methods is essential for the field operator, policy decisions must be made by responsible professionals. However, the administrator who is responsible for control action does not have access to academic training (as is the case in most natural-resource fields). Yet, each day the problems he must deal with become more complex.

Pest control workers often complain that their jobs lack prestige. But respect and recognition will come from the positive results of scientifically sound control programs, rather than from inducing well-known biologists or administrators to attend pest control meetings (Howard, 1962).

The public tends to be impressed by the sheer numbers of pest animals that have been destroyed. Hence, control efforts achieving less than spectacular results may be difficult to sell. Value judgments by administrators or control specialists in this area should not be biased by economic, social, or political considerations. The ability to make judgments without such bias is evidence of education.

ETHICS AND PEST CONTROL

For most of human history, man has been unable to achieve control over wild animals. They often threatened his physical and economic survival; hence the development of his belligerent attitude toward such animals. Man's recent spectacular population growth and proliferating technology cause him now to overrun many of the forces in his environment that he once could not master.

In an appraisal of rare and declining birds of the world, Vincent (1966) listed 646 forms. The ultimate survival of some 260 of these is threatened by the activities of man (Table 1). The issue of control versus bird conservation must be reconciled by balancing utilitarian and ethical values.

We now realize that to control an animal population, of whatever size, it

TABLE 1   Rare and Vanishing Birds Threatened by the Activities of Man

Principal Cause of Threat	Number of Forms	Example
Direct predation	85	Eskimo curlew (*Numenius borealis*)
Zoo and aviculture collecting	5	Monkey-eating eagle (*Pithecophaga jefferyi*)
Disturbance	17	California condor (*Gymnogyps californianus*)
Pesticides	4	Japanese white stork (*Ciconia ciconia*)
Introduced disease	3	Maui parrot bill (*Pseudonestor xanthophrys*)
Introduced predators	47	Jamaican black rail (*Laterallus jamaicensis*)
Habitat destruction	103	Ivory-billed woodpecker (*Campephilus principalis*)

Source: Vincent (1966).

is necessary to alter those facets of its ecology, such as time and space relationships, food and cover relationships, or competitor and predator relationships, so as to render the population unstable. Thus manipulated, the unstable population becomes more vulnerable to its limiting factors. The technique must be selected on the dual standards of utility and ethics, and the results appraised on those standards.

Because an animal becomes a pest only in the context of a given time and place, indiscriminate mass destruction is expensive, pointless, and often wasteful. When control of relatively rare animals is necessary, the general or regional survival of the species must be ensured, for example, by creating a sanctuary. Destruction of a natural habitat to control pests or to provide opportunity for commercial or economic development must be undertaken very carefully, or neither objective will be achieved. Indeed, very often the wisest use of land is to leave it untouched.

The dangers inherent in certain kinds of pest control may have not yet materialized, but this does not justify proceeding as if there were no danger. Side effects, residual effects, and delayed reaction to chemical control are not fully understood, nor are data complete. Until latent effects are fully appraised, it is well to move cautiously.

Tradition, habit, and ethnic culture frequently dictate how a pest may be controlled. In no case, however, should the ecological approach be lost sight of. Moreover, while it might be economically profitable to export pesticides that do not meet standards set for domestic use by the U.S. Food and Drug

Administration, it is both unethical and politically hazardous to do so.

To many laymen and some professionals, control means killing with poisons. A recent report to the U.S. Department of the Interior (Leopold, 1964) recommended that only those methods of control that are efficient, selective, safe, humane, and economical should be considered. Where a control method meeting all the stated criteria does not exist, public and private efforts should develop one. In general, quick-acting lethal poisons are difficult to handle safely or selectively in the field. For example, the popular mammal poison 1080 (sodium fluoroacetate) would not meet the above criteria. Leopold (1964) therefore recommends that 1080 be prohibited as a field rodenticide and that exports be stopped to foreign countries having inadequate safeguards.

The use of lethal poisons is apparently accepted by even the most squeamish individuals. The success of poisons in controlling insects conditions people to using them against other species. Yet it cannot be assumed that all lethal poisons are humane, safe, or selective, however effective and economical they may be. Poisons should be used only as a temporary check in acute pest problems, and only after all else fails.

Bounty systems, as discussed previously, have long been used in attempts to control predators and disease carriers. Whereas short-range, limited control may be achieved, wide-scale killing for bounty does not effect overall control; it is deceptive and expensive. The biological, economic, and administrative aspects of the bounty cannot be justified as protecting the public. The basic justification of pest control should be that the methods used and the results obtained are ecologically sound, socially acceptable, and minimally harmful to the biota, including man.

REGULATORY ASPECTS OF ANIMAL CONTROL

The increasing use of chemicals for the control of vertebrate pests has resulted in broadening the regulations of the federal Insecticide, Fungicide, and Rodenticide Act of 1947 to include new products appearing on the market. The word *pest* was redefined to include certain birds, predatory animals, poisonous reptiles, rough fish, plant pathogens, and weeds.

The Act introduced a registration concept intended to ensure the safety and efficacy of a material before it is marketed. The legislation was oriented to both producer and consumer and was designed to protect the public physically and economically. It required that all pesticides moving interstate meet U.S. Department of Agriculture (USDA) standards for effectiveness and safety. When they meet these standards, they are "registered," a status that may be cancelled at any time if subsequent evidence so indicates. Registration does not imply recommendation. The product label must show its chemical composition; the specific pests it will control; the crops, if any, on which it may be used; how it may be applied safely and effectively; and precautions.

Throughout the country, federal (USDA) inspectors and cooperating state enforcement officials check unregistered pesticides and products suspected of being misbranded or adulterated. When tests show a product to be below standards, it is taken off the market and legal action is initiated against the offender (Anderson, 1966a).

A uniform state insecticide, fungicide, and rodenticide act has been developed by the Council of State Governments and has been modified to incorporate 1959 and 1964 amendments to the federal Insecticide, Fungicide, and Rodenticide Act.

So far as human welfare is concerned, much is known about the short-term and acute health hazards of pesticides, but comparatively little is known about the dangers from long-term exposure to minute quantities in the environment or about the dangers to persons who are regularly exposed as repeated users of pesticides. Recognition of these unsolved problems by government agencies led to the establishment of the Pest Control Review Board in 1961. The Board soon demonstrated that pest control could be made safer. In 1964, the Board was expanded and became the Federal Committee on Pest Control, with greater responsibilities. The Committee is composed of eight members, appointed by the Secretary of Agriculture, the Secretary of Defense, the Secretary of the Interior, and the Secretary of Health, Education, and Welfare (two appointments by each secretary). In addition, there are liaison members (nonvoting) from the National Science Foundation and the Department of State, the Department of Transportation, and the Department of Housing and Urban Development. The Committee reviews any pest-control activity that involves the use of federal funds; coordinates federal pesticide research, pesticide monitoring, and public information programs; and evaluates the beneficial and harmful effects of pesticide use before a control program is initiated. It has no executive powers and can only recommend administrative action or propose legislation, but it has been effective in winning acceptance of its recommendations.

The Federal Committee on Pest Control also suggests a pattern for cooperative effort at state and local levels, because private users on farms and in forests, gardens, and homes apply about 95 percent of the pesticides used in the United States. As Anderson (1966b) indicated, the problem is acute, since there are no control bodies in about 40 percent of the states.

The legal situation relating to welfare of wildlife is another matter. Flexibility in the laws may allow otherwise protected animal species to be destroyed when they conflict with human interests. Laws vary from one state to another, as do their enforcement and penalty provisions.

Control procedures harmful to wildlife other than target animals should not be used. In most cases there are alternative treatments, although they may be more expensive or less efficient. Tradition often prevents consideration of the alternatives. The federal Insecticide, Fungicide, and Rodenticide Act would be more effective if it included a provision that would automatically prohibit the use of any chemical highly toxic to many species of fish and other wildlife whenever a less hazardous chemical is available.

Certain pesticides should be restricted to use by professionals only, as are many of our drugs. (Some of the more potent chemicals that might fit into this restricted group are 1080, parathion, endrin, and thallium.) The safe and effective use of commercial poisons involves more than the chemical itself; it involves the method of formulation and application. Two or more chemicals are sometimes employed, thereby increasing the hazard. But laws alone cannot eliminate the misuse of commercial poisons. Thus, the federal government has emphasized educating the user, on whose judgment wise use depends.

What of abuses at the field level? It is often assumed that government predator-control agents are the offenders but private users are guilty also, possibly to a greater extent. It is therefore necessary that regulation of activities in the field as well as in the laboratory be undertaken by government. Ultimately, local legislation making the user responsible for the misapplication of pesticides may be required.

REFERENCES

Anderson, R. J. 1966a. Official registration of pesticides, pp. 388–391. *In* Scientific aspects of pest control, NAS–NRC Publ. 1402. National Academy of Sciences, Washington, D.C.

Anderson, R. J. 1966b. The federal committee on pest control, pp. 367–374. *In* Scientific aspects of pest control, NAS–NRC Publ. 1402. National Academy of Sciences, Washington, D.C.

Howard, W. E. 1962. Means of improving the status of vertebrate pest control. Trans. N. Amer. Wild. and Natur. Resour. Conf. 27:139–150.

Jackson, W. B. 1964. Undergraduate preparation, pp. 122–126. *In* Proc. 2nd Bird Control Seminar, Bowling Green State Univ., Bowling Green, Ohio.

Leopold, A. S. (Committee Chairman). 1964. Predator and rodent control in the United States. Trans. N. Amer. Wildl. and Natur. Resour. Conf. 29:27–49.

Vincent, J. (compiler). 1966. Red data book. Vol. 2, Aves. Intl. Union for Conserv. of Nature and Natur. Resour., Morges, Switzerland, 275 pp.

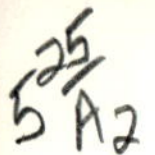